CAMBRIDGE SCIENCE CLASSICS

THE THEORY OF HOMOGENEOUS TURBULENCE

THE THEORY OF HOMOGENEOUS TURBULENCE

BY

G. K. BATCHELOR

Professor of Applied Mathematics,
University of Cambridge

CAMBRIDGE
UNIVERSITY PRESS

CAMBRIDGE UNIVERSITY PRESS
Cambridge, New York, Melbourne, Madrid, Cape Town, Singapore,
São Paulo, Delhi, Dubai, Tokyo, Mexico City

Cambridge University Press
The Edinburgh Building, Cambridge CB2 8RU, UK

Published in the United States of America by
Cambridge University Press, New York

www.cambridge.org
Information on this title: www.cambridge.org/9780521041171

First published, in the Cambridge Monographs on
Mechanics and Applied Mathematics series, 1953
Reprinted 1956, 1959
Reissued in the Cambridge Science Classics series 1982
Reprinted 1986, 1990, 1993

A catalogue record for this publication is available from the British Library

Library of Congress catalogue card number: 82–4373

ISBN 978-0-521-04117-1 Paperback

To

G.I.

CONTENTS

VI The Universal Equilibrium Theory

VII Decay of the Energy-containing Eddies

VIII The Probability Distribution of $\mathbf{u}(\mathbf{x})$

PREFACE

Great advances in the theory of turbulent motion have been made in the years since the war, and although much remains to be done it seems an opportune time to put some of the existing work into more permanent form. There are many excellent reviews in the periodical literature, but they are necessarily limited in size and have usually been concerned with recent advances. It has been my experience that students and others coming fresh to the subject have found a need for a systematic and complete account of the established knowledge of turbulence, and I hope that this book will fill part of that need. There is perhaps another reason why the present time is suitable for the appearance of a book on homogeneous turbulence. It seems that the surge of progress which began immediately after the war has now largely spent itself, and there are signs of a temporary dearth of new ideas. Work on turbulence in flows with a mean rate of shear is proceeding apace, but in the more refined analysis of homogeneous turbulence we have got down to the bedrock difficulty of solving non-linear partial differential equations. Many properties of homogeneous turbulence can be deduced without a quantitative treatment of the non-linear aspects of the Navier-Stokes equation, but it is probable that not many of significance remain to be found in this way. Such a treatment will not be developed overnight.

I have tried to make the following account of homogeneous turbulence as complete as possible, with one qualification, mentioned below, and I hope no significant work has been overlooked. In the selection of experimental data and in the assessment of theoretical contributions it will be found that I have leaned heavily on the research carried out at Cambridge. If in places there seems to be a bias unsupported by objective arguments, I can only apologize and promise to consider all suggestions for improvement.

The exception to completeness is in the exclusion of all work requiring a Lagrangian, or motion-of-a-particle, description of the flow. Thus no mention of the important problem of turbulent diffusion will be found. The methods used to handle Eulerian and Lagrangian descriptions of the motion are very different and they proceed independently, so that problems like diffusion would not fit naturally into the present work. Moreover, there is a considerable amount of research now in progress on problems of particle motion and we may know a good deal more in a few years' time.

Finally, it may be worth while to say a word about the attitude that I have adopted to the problem of turbulent motion, since workers in the field range over the whole spectrum from the purest of pure mathematicians to the most cautious of experimenters. It is my belief that applied mathematics, or theoretical physics, is a science in its own right, and is neither a watered-down version of pure mathematics nor a prim form of physics. The problem of turbulence falls within the province of this subject, since it is capable of being formulated precisely. The manner of presentation of the material in this book has been chosen, not with an eye to the needs of mathematicians or physicists or any other class of people, but according to what is best suited, in my opinion, to the task of *understanding the phenomenon*. Where mathematical analysis contributes to that end, I have used it as fully as I have been able, and equally I have not hesitated to talk in descriptive physical terms where mathematics seems to hinder the understanding. Such a plan will not suit everybody's taste, but it is consistent with my view of the nature of the subject-matter.

A first draft of this work was one of the essays to which the Adams Prize for 1949–50 in the University of Cambridge was awarded.

It is a pleasure to be able to record here my gratitude to Sir Geoffrey Taylor, Dr A. A. Townsend and Professor W. Heisenberg, who have contributed much to my understanding of turbulence; the many conversations I have had with them are among my happiest experiences. I am grateful also to Mr T. H. Ellison, Dr I. Proudman and Dr R. W. Stewart, who were students of turbulence at Cambridge during the period of preparation of the book, for the help they gave in many different ways; to Professor M. S. Bartlett, Dr M. Mitchner and Mr J. E. Moyal for giving me their comments on parts of the manuscript; and to my wife for many hours spent in typing.

Acknowledgement is made to the following bodies for permission to reproduce figures from their publications: the Royal Society (figs. 5.1, 5.2, 6.3, 7.1–7.9, 8.5), the Cambridge Philosophical Society (figs. 6.3, 7.10, 7.11, 8.3, 8.6), *Quarterly Journal of Mechanics and Applied Mathematics* (figs. 4.1, 4.2, 5.3), and the Institute of Aeronautical Sciences (fig. 6.2).

G. K. B.

CAMBRIDGE
July 1952

INTRODUCTION

1.1. The study of homogeneous turbulence

The problem to be studied herein can be stated very briefly, provided we do not worry about details for the moment. We imagine an infinite uniform body of fluid which can be characterized in the usual way by a density ρ and molecular transport coefficients such as the viscosity μ. This body of fluid can be set into different kinds of motion, many of which are described in the textbooks on hydrodynamics. It is a well-known fact that under suitable conditions, which normally amount to a requirement that the kinematic viscosity ν be sufficiently small, some of these motions are such that the velocity at any given time and position in the fluid is not found to be the same when it is measured several times under seemingly identical conditions. In these motions the velocity takes random values which are not determined by the ostensible, or controllable, or 'macroscopic', data of the flow, although we believe that the *average* properties of the motion are determined uniquely by the data. Fluctuating motions of this kind are said to be turbulent. Our concern is with homogeneous turbulence, which is a random motion whose average properties are independent of position in the fluid. The problem is to understand the mechanics, and to determine analytically the average properties, of this kind of motion.

The conception of homogeneous turbulence is idealized, in that there is no known method of realizing such a motion exactly. The various methods of producing turbulent motion in a laboratory or in nature all involve discrimination between different parts of the fluid, so that the average properties of the motion depend on position. However, in certain circumstances this departure from exact independence of position can be made very small, and it is possible to get a close approximation to homogeneous turbulence. It has been found, for instance, that if a uniform stream of fluid passes through a regular array of holes in a rigid sheet, or a regular grid of bars, held at right angles to the stream, the motion downstream of the sheet consists of the same uniform velocity together with a superimposed random distribution of velocity. This random motion dies away with distance from the grid, and to that extent is not statistically homogeneous, but the rate of decay is found to be so small that the assumption of homogeneity of the turbulence is valid for most purposes. Thus there is available a convenient laboratory method of producing turbulence which is approximately homogeneous, the various stages of decay occurring at different distances from the grid, and to this fact, primarily, must be attributed the very considerable advances in the subject in recent years. The possibility of carrying out controlled and accurate experiments rapidly has permitted a very fruitful combination of experimental and theoretical research.

The kinds of turbulent motion which are encountered in nature or in the fields of aeronautics, hydraulics and chemical engineering are usually more complicated than homogeneous turbulence. These turbulent motions usually are such that, in the first place, there is a variation of the mean velocity with position (which normally arises from the presence of rigid boundaries) and, in the second place, there is a variation of the average properties of the turbulent, or fluctuating, velocity with position. As a consequence of these two properties there will occur some kind of interaction between the fluctuating and mean components of the motion—or, stated in another way, the presence of rigid boundaries imposes a steady boundary condition on the random velocity field—which is difficult to handle mathematically, and there will also be transport effects produced by the different intensity of the fluctuating motion at different points. These are complicated mechanical effects and we have not yet obtained a proper understanding of them. As a preliminary, it seems appropriate to consider homogeneous turbulence which has neither of the two properties mentioned above. Despite the lack of a wide field of immediate application of the results concerning homogeneous turbulence, the study of homogeneous turbulence has practical utility, in that if we understand this simpler case we also understand some at least of the aspects of non-homogeneous turbulence.

It will be clear that if we wish to go further in the direction of simplifying the problem in order to make it more tractable, we can make assumptions about the directional symmetry of the average properties of the turbulent motion. In the simplest possible case the turbulence is statistically homogeneous and isotropic, and so depends on neither the position nor the direction of the axes of reference. The possibility of this further assumption of isotropy exists only when the turbulence is already homogeneous, for certain directions would be preferred by a lack of homogeneity. It has been found that, in addition to being the simplest possible case of turbulent motion, isotropic turbulence is readily generated in the laboratory. Whatever the initial directional properties of a field of homogeneous turbulence, it appears to settle down to an approximately isotropic state, and the usual laboratory method of generating homogeneous turbulence by passing a uniform stream through a regular array of bars produces, in fact, turbulence which is very nearly isotropic. Consequently, most of the available data concerns isotropic turbulence. Equally, most of the theoretical work has been built on this foundation, so that the greater part of this book will be devoted to the special case of isotropic turbulence. However, it will sometimes be found possible to retain greater generality, and there are some definite and important results for non-isotropic homogeneous turbulence.

Finally, if we are to list the reasons for studying homogeneous turbulence, we should add that it is a profoundly interesting physical phenomenon which still defies satisfactory mathematical analysis; this is, of course, the most compelling reason.

1.2. Mathematical formulation of the problem

Let us consider first the equations which determine the variation of the turbulent velocity with respect to position and time.

One governing equation is provided by the so-called continuity equation expressing the conservation of mass of the fluid. In the general case it is necessary to allow for variations in the density of the fluid in space and time consequent upon variations in the pressure. A measure of the importance of these density variations is provided by the ratio of a typical fluid velocity (say, in our case, the root-mean-square) to the average velocity of sound in the fluid;†

† See, for instance, Chap. 1 of *Modern Developments in Fluid Dynamics*, vol. 3 (High-Speed Flow), Oxford University Press, 1953.

when this ratio is small compared with unity the density variations are negligible and the fluid behaves as though it were incompressible. It is possible that in certain astrophysical situations the turbulent velocities are comparable with the velocity of sound. However, in the cases of turbulent motion set up in the laboratory or observed in the atmosphere, oceans or rivers, it is almost certain that this ratio will be very small compared with unity for the reason that the amount of kinetic energy involved would otherwise be prohibitively large.

Consequently we shall confine attention to the case of a fluid which is effectively incompressible (indeed, it would be difficult to proceed on any other basis on account of the complexity of the problem). The equation of continuity is then

$$\nabla . \mathbf{u} = 0, \qquad (1.2.1)$$

where $\mathbf{u}$ is the vector velocity of the turbulent motion at a position in the field specified by the vector coordinate $\mathbf{x}$, where both $\mathbf{u}$ and $\mathbf{x}$ are referred to axes such that the fluid has no average motion.

Further information is necessary to specify the variation, if any, of the density ρ and kinematic viscosity $\nu\,(=\mu/\rho)$. The simplest case is clearly that of an isothermal fluid (which is free from external force) for which ρ and ν are constant with respect to position and time, and this case also corresponds to that normally encountered in the laboratory or, provided the scale is small enough, in the atmosphere. We shall therefore assume ρ and ν to be uniform.

As in other branches of fluid dynamics, we shall assume that the Navier-Stokes equation of motion is valid. The variation of $\mathbf{u}$ with $\mathbf{x}$ and time t then satisfies

$$\frac{\partial \mathbf{u}}{\partial t} + \mathbf{u} . \nabla \mathbf{u} = -\frac{1}{\rho} \nabla p + \nu \nabla^2 \mathbf{u}, \qquad (1.2.2)$$

where ∇ represents the gradient operator with respect to the coordinate system $\mathbf{x}$, and p represents pressure. The validity of this equation will be taken as a fundamental premise; the *a priori* justification is that we see no reason why the medium should not behave as a fluid, characterized by the parameters ρ and ν and by differentiable functions $\mathbf{u}$ and p, for turbulent motions equally as for the many non-turbulent motions for which its validity has been amply confirmed by experiment. Confirmation that the adoption of (1.2.2) leads to predictions in agreement with observations of turbulent motion has also been obtained and will be described later. It has occasionally been speculated in the literature of the subject that since turbulence is a mixture of many different subsidiary motions having different length scales, there may exist some subsidiary motions whose length scales are so small as to be comparable with the mean free path (for a gaseous medium), and if this were true the Navier-Stokes equation certainly would not apply to these small-scale motions. However, the action of viscosity is to suppress strongly the small-scale components of the turbulence, and we shall see that for all practical conditions the spectral distribution of energy dies away effectively to zero long before length scales comparable with the mean free path are reached. As a consequence we can ignore the molecular structure of the medium and regard it as a continuous fluid.

The set of equations (1.2.1) and (1.2.2) is now sufficient to determine $\mathbf{u}$ and p as functions of $\mathbf{x}$ and t when the boundary conditions (with respect to both $\mathbf{x}$ and t) are specified. In a problem of non-turbulent, or laminar, motion the boundary conditions are specified definitely, and, provided the boundary conditions are complete, a definite and (usually) unique solution can be obtained, at least in principle. In some cases of turbulent motion the appropriate conditions are as for the corresponding laminar motion together with a perturbation of the velocity distribution. For instance, the classic Reynolds problem of flow which enters a tube smoothly, but with a small perturbation, and becomes turbulent farther down the tube, could be formulated in this way. In such cases the turbulence develops from the perturbation by means of a transfer of energy from the (initially) smooth flow, and the exact nature of the perturbation is not important. In other cases, among which is homogeneous turbulence, turbulent motion is given as part of the initial conditions and the problem is to follow its history.

In the case of homogeneous turbulence the boundary conditions with respect to $\mathbf{x}$ are specified, in effect, by the statistical uniformity of the motion with respect to position and need no further consideration. (It is, of course, necessary that the fluid extend to infinity in all directions.) The boundary conditions with respect to t are that at some initial instant the velocity is a random function of position conforming to given probability laws. (The method of specifying statistically a random function of position will be considered in Chapter II.) In practice there is not much chance that this initial statistical information would actually be known, but it needs to be known in an idealized determinate mathematical formulation of the problem.

It is clear that if the initial conditions of the turbulent motion are known in probability only, we cannot hope to do more than determine the velocity field at later instants in the same way; nor, of course, should we wish to do so, any more than we should wish to determine the instantaneous positions and velocities of the molecules of a gas. It has not yet been established rigorously that it is possible to do even as much as this; however, the consensus of opinion is that if a random velocity field is specified statistically at one instant and if rules (viz., in our case, equations (1.2.1) and (1.2.2)) are given for determining the way in which any particular velocity distribution changes with time, then the subsequent random velocity field is statistically determinate. Hence the mathematical formulation of the problem of homogeneous turbulence is this: *Given an infinite body of uniform fluid in which motions conform to the equations* (1.2.1) *and* (1.2.2), *and given that at some initial instant the velocity of the fluid is a random function of position described by certain probability laws which are independent of position, to determine the probability laws that describe the motion of the fluid at subsequent times.*

Now the class of probability laws specifying a random infinite field of velocity is very wide, and we are not interested in all of them as possible initial conditions of the turbulent motion. For instance, we should not be interested in those which permitted discontinuities in the spatial distribution of velocity. However, even when such obviously unreal cases are excluded, the range of variation of the initial statistical conditions is too broad for the investigation of particular sets of initial conditions to be of significance (even supposing it to be possible mathematically). Instead, we put our faith in the tendency for dynamical systems with a large number of degrees of freedom, and with coupling between these degrees of freedom, to approach a statistical state which is *independent* (partially, if not wholly) of the initial conditions. With this general property of dynamical systems in mind, rather than investigate the motion consequent upon a particular set of initial conditions, we explore the existence of solutions which are asymptotic in the sense that the further passage of time changes them in some simple way only. Since the energy of turbulent motion is being dissipated by viscosity continually, we cannot have the simple situation of the kinetic theory of gases, in which the asymptotic statistical state of the molecular motion is independent of time. The elucidation of the kind of asymptotic statistical state to be expected is the crux of the problem of homogeneous turbulence, and we shall have more to say about it in later chapters. Meanwhile it should be kept in mind that the general method of attacking the problem as formulated at the end of the preceding paragraph is indirect, inasmuch as we attempt to guess the ultimate statistical state of the turbulence and to show that this statistical state would follow from a whole class of different initial conditions.†

1.3. Brief history of the subject

It will perhaps be useful, as an introduction to the work to be described in the following chapters, to recount briefly the major contributions to the subject made over the last fifteen years. We shall see later that the principal difficulties in the way of a solution of the problem of homogeneous turbulence arise from:

(*a*) the three-dimensional character of the velocity field,

(*b*) the non-linearity of the equation of motion, and

(*c*) the random variation of the velocity and the need for statistical methods.

† There is clearly a close correspondence between the subjects of turbulence and statistical mechanics. Statistical mechanics is concerned with the properties of a large number of particles whose motion conforms to certain collision laws, whereas turbulence is concerned with a continuum whose motion conforms to (1.2.1) and (1.2.2). In both cases the interest lies in asymptotic statistical states which are independent of a wide class of initial states. One might also say that statistical mechanics shows how comprehensive and far-reaching are the deductions made on the basis of the particle structure of matter, irrespective of the nature of these particles and the precise form of the laws of their interaction. It is a very striking fact that some of the recent work in homogeneous turbulence (see Chapters VI–VIII) is tending to ignore the precise form of the equations (1.2.1) and (1.2.2) and to make use of their general features (e.g. dissipation of energy, non-linearity) only.

These three difficulties occur in an interconnected fashion and are not to be overcome quite separately, but the history of the research carried out on the problem is largely an account of contributions to one or other of the three different aspects.

The origin of the subject lies in G. I. Taylor's pioneering work in 1935 (Taylor, 1935*a*). Prior to this time there had been no clear recognition and acceptance—save in an earlier paper by Taylor himself (Taylor, 1921)—of the fact that the velocity of the fluid in turbulent motion is a random continuous function of position and time, and theories of turbulence were based on analogies with the discontinuous collisions between discrete entities that have been studied in the kinetic theory of gases. Taylor broke with these primitive concepts and introduced the correlation between the velocities at two points as one of the quantities needed to describe the turbulence. As soon as the statistics of continuous random functions were considered, it became clear that the assumption of statistical homogeneity would greatly simplify the analysis; Taylor went further still and considered isotropic turbulence. In this same paper Taylor described measurements which showed that the turbulence generated downstream from a regular array of rods in a wind tunnel was approximately homogeneous and isotropic. Thus a clear guide to the opportunities for further theoretical and experimental work was established.

Further important contributions to the subject were made by Taylor in 1938. The first was a consideration of the mechanical processes represented by the non-linear term (found to be related to the mean value of the product of three velocities) in the equation for the decay of mean-square vorticity (Taylor, 1938*a*). This work demonstrated clearly two important consequences of the non-linearity of the dynamical equation: the skewness of the probability distribution of the difference between the velocities at two points, and the existence of an interaction or modulation between components of the turbulence having different length scales. The second contribution was the introduction into turbulence theory of a result obtained in pure mathematics, viz. that the Fourier transform of the correlation between two velocities is an energy spectrum function in the sense that it describes the distribution of kinetic energy over the various Fourier wave-number components of the turbulence

(Taylor, 1938*b*). This contribution to aspect (*c*) of the problem has been developed further in more recent years.

Soon after Taylor's work T. von Kármán perceived that mean values of the products of the velocities at two (or more points) were tensors, which immediately enabled the analysis to be expressed more concisely and greatly facilitated the deductions from the assumption of isotropy (v. Kármán, 1937*a*, *b*). In particular, v. Kármán and Howarth (1938) were able to show that mean values of the product of two and of three of the components of the velocities at two points could each be expressed in terms of a single scalar function when the turbulence is isotropic; the Navier-Stokes equation then provides a differential relation between these two functions, the independent variables being the time of decay and the distance between the two points at which the velocities are taken. (The idea of using the Navier-Stokes equation to relate mean velocity products of different orders was first advanced by L. Keller and A. Friedmann (1924); however, without the simplification of homogeneity and isotropy it was not possible to proceed far.) Another big advance in the development of the kinematics of mean values of velocity products in isotropic turbulence came when H. P. Robertson (1940) showed how an isotropic tensor of arbitrary order could be expressed in terms of the known invariants of the rotation group. The same methods have been used to analyse the kinematics of axisymmetric turbulence (Batchelor, 1946; Chandrasekhar, 1950*a*). All this work is, in effect, an attack on aspect (*a*) of the problem.

The first attempts to handle the dynamical problem systematically were made by v. Kármán (1937*a*, *b*, 1938) who introduced, for the purposes of simplification, the assumption of 'self-preservation' of the shape of the velocity product functions during decay. This reduced the number of independent variables in the dynamical equation to one, but the equation still contained two dependent variables (viz. velocity products of order n and $n+1$), so that definite results could only be obtained under suitably restricted conditions. There was available some experimental evidence that the correlation functions did not preserve their shape over the whole of the range of the distance between the two points, and it gradually became clear that the assumption of self-preservation, or similarity during decay, while a useful mathematical tool, would need a sound physical basis if its limitations were to be understood.

The required physical basis for one kind of similarity of the turbulence was suggested some years later by A. N. Kolmogoroff (1941 *a*, *c*). This work was overlooked in countries other than the U.S.S.R. for several years, but its profound importance has since been appreciated and it is now the starting-point for many researches. Kolmogoroff's hypothesis was that the small-scale components of the turbulence are approximately in statistical equilibrium. These small-scale components owe their existence to the non-linear interchange of energy between different wave-number components, and Kolmogoroff postulated that the equilibrium would be universal, apart from the effect of variation of two parameters, one the viscosity of the fluid and the other determined by the large-scale components of the turbulence. Thus, when these two parameters are given, the complete statistical specification of the small-scale components of the turbulence is determined, and many definite predictions may be made from dimensional analysis. Experimental support for the theory was not available immediately, owing to the very small size of the components to which the theory applies at the Reynolds numbers ordinarily used in the laboratory, but there is now sufficient evidence to warrant confidence in the theory. The idea of a statistical equilibrium of the small-scale components was also put forward, independently, by A. Obukhoff (1941), by L. Onsager (1945), and by C. F. von Weizsäcker (1948).

Much of the work stemming from the equilibrium theory has been concerned with the application to special situations, such as turbulent shear flow, atmospheric diffusion, turbulence in an electrically conducting fluid, in which the universality of the statistical equilibrium provides useful information. Other work has been devoted to a more detailed study of the equilibrium, e.g. to a consideration of the distribution of energy over the wave-numbers within the equilibrium range. For this purpose it is necessary to make some assumption about the non-linear transfer of energy across the spectrum. An assumption of this kind was introduced by W. Heisenberg (1948 *a*), but it is not wholly convincing, and recent measurements show that it does not lead always to accurate results. The determination of the transfer of energy across the spectrum produced by the non-linear inertia forces is still the central difficulty of the problem of turbulent motion.

An alternative plan for investigating the effect of the non-linear inertia forces has been followed by J. M. Burgers over a number of years beginning in 1939 (for a review, see Burgers, 1948*a*). Burgers has explored the solutions of systems of equations which are simpler than (1.2.1) and (1.2.2) but which retain the essential features of the non-linear term and the viscosity term. In particular, he has explored the tendency for the non-linear terms to produce steeper gradients of velocity as time proceeds, and the tendency for viscous forces to damp out rapid changes of velocity, the two effects together leading to the permanent existence of 'dissipation layers' in which there is a high rate of viscous dissipation of energy. Some of the results obtained for the model system are strikingly similar to those obtained experimentally, and Burgers believes that investigation of the model system will suggest methods of attacking the full Navier-Stokes equation. However, it has not been found possible to obtain solutions of the model system which are random functions of position and time, so that the link between the model system and reality is not yet close.†

On the experimental side, a large number of measurements of mean values of different velocity products has been made since 1935, principally with the hot-wire anemometer. This instrument is capable of giving an electrical signal which is proportional to the instantaneous velocity of the fluid at the point where the wire is placed, and a great deal of skill has been used to devise electrical circuits that will analyse the output signal in such a way as to give some particular mean value. These developments in technique will not be described in this book, but the importance of the part which they have played in the research should not be forgotten, and the fact that some knowledge of the errors involved in the different hot-wire measurements is necessary for a proper appraisal of the validity of theoretical predictions should be kept in mind. Examples of the kinds of measurement of which the hot-wire anemometer is capable are to be found in papers by A. A. Townsend (e.g. 1947),

† Note added in proof: But see a recent paper by J. M. Burgers, 'Sur un modèle simplifié de la turbulence', *Publ. Sci. Tech. Ministère l'Air*, no. 251, 1951.

who has supplied many of the measurements on which our present ideas about homogeneous turbulence are based, and S. Corrsin (1947).

One of the first measurements to be made was of the decay of kinetic energy as a function of time, and it has been repeated many times under different conditions (Dryden, 1941; Batchelor and Townsend, 1948*a*; Stewart and Townsend, 1951). There seems now to be general agreement that the decay proceeds according to a very simple law for an initial period after the formation of the turbulence, viz. that the kinetic energy (per unit volume of fluid) is inversely proportional to the lapse of time since the (virtual) instant at which the energy would have been infinite. There have been several attempts to explain this decay law but none is completely satisfactory. An explanation which seems promising is that this decay law is one of the properties of an asymptotic statistical state of the larger, energy-containing components of the turbulence; it is postulated that these large-scale components of the turbulence are in a kind of quasi-equilibrium, which, owing to the decay in time, is one degree less universal than the absolute equilibrium of the small-scale components (Heisenberg, 1948*b*). A difficulty with this theory is that the correlation functions do not seem to change with time in the simple self-similar manner that would be expected from the quasi-equilibrium. Also it is not easy to account for the termination of this initial period of decay. However, these developments are much too recent to be written up as 'history', and they will be described in their proper place later.

After a sufficiently long time of decay, the energy of the turbulence is found to decrease as the $(-\frac{5}{2})$ power of the time. This has been explained (Batchelor and Townsend, 1948*b*) as one of the consequences of the non-linear inertia terms being negligible. In this final period of decay the velocity satisfies a heat-conduction type of equation, and a complete determination of the statistical parameters describing the motion at any time can be made.

Simultaneously with all the above work there has been a steady development of the purely mathematical problems associated with stationary random functions, and the mathematical results have gradually been introduced and exploited in turbulence theory. Taylor's use of the spectrum function which is the one-dimensional

Fourier transform of the velocity correlation has already been mentioned. Later J. Kampé de Fériet (1948) and G. K. Batchelor (1949*a*) introduced the three-dimensional spectrum function, which provides a suitable tool for the application of the various similarity hypotheses, and investigated many of its exact properties. Heisenberg (1948*a*) had already achieved the same end, in effect, by working heuristically in terms of the Fourier coefficient obtained from a three-dimensional analysis of the velocity field. Three-dimensional Fourier series were also used by G. Darrieus (1938) in a paper which anticipated much of the later work. On the whole it seems probable that the mathematical techniques devised to cope with the three-dimensional character of the velocity field are not far from their optimum form. However, many of the mathematical difficulties arising from the random character of the velocity are not yet fully resolved.

There are very many isolated papers and incomplete lines of research which cannot appropriately be mentioned here but will find a place in later chapters.

CHAPTER II

MATHEMATICAL REPRESENTATION OF THE FIELD OF TURBULENCE

2.1. Method of taking averages

Since we are supposing, as a matter of definition, that the velocity in a turbulent flow takes random values, and since, as a consequence, we are interested only in the *average* values of quantities, it is necessary to lay down rules about the method of taking averages.

What do we mean when we say that the velocity **u** is a random function of position **x** and time *t*? We mean that at a given point (**x**, *t*) of space-time the velocity **u** is not predictable from the data of the problem but takes random values; it is an implication of the use of the phrase 'random function' that these values are distributed according to certain definite probability laws. We shall assume that the probability laws describing the velocity fluctuations *are* determined by the data of the problem. If we perform the (idealized) experiment of generating a field of turbulence in some way, say by sweeping a grid of bars through still air, and subsequently allowing it to decay, we can measure the value of the velocity at the given point. (Such an experiment, which supplies a value of **u** for every point (**x**, *t*), will be termed a *realization* of the field of turbulence.) The same experiment repeated many times supplies just as many different values of **u**, and we are defining the turbulent motion to be such that as the number of experiments tends to infinity the frequency distribution of **u** tends to a limiting form which is determined by the ostensible data of the problem (i.e. by the shape of the grid of bars, etc.).

This interpretation of **u** as a random variable applies for every point (**x**, *t*), and in general there will exist a statistical connexion between the random values of **u** that occur at different points in space-time. The specification of this statistical connexion will be included in the joint-probability distribution of the values of **u** at the chosen points in space-time, and there will be one such distribution for every different set of points in space-time. All these joint-probability distributions are a part of the specification of the random function, and are determined (or so we shall assume) by the ostensible data of the problem; if we wish we can also regard them as measurable from a large number of realizations of the field.

We can now define a mathematical expectation, or probability average, or ensemble average, by reference to the joint-probability distributions of the values of $\mathbf{u}$ at different sets of points in space-time. If $F(\mathbf{u}_1, \mathbf{u}_2, \ldots, \mathbf{u}_n)$ is some function of the values $\mathbf{u}_1, \mathbf{u}_2, \ldots, \mathbf{u}_n$ which the velocity takes at the points $(\mathbf{x}_1, t_1)$, $(\mathbf{x}_2, t_2)$, ..., $(\mathbf{x}_n, t_n)$ respectively, F is a random variable and its probability average is defined as

$$\bar{F} = \int F\, P(\mathbf{u}_1, \mathbf{u}_2, \ldots, \mathbf{u}_n)\, d\mathbf{u}_1\, d\mathbf{u}_2 \ldots d\mathbf{u}_n, \qquad (2.1.1)$$

where $P(\mathbf{u}_1, \mathbf{u}_2, \ldots, \mathbf{u}_n)$ is the joint-probability density function of $\mathbf{u}_1, \mathbf{u}_2, \ldots, \mathbf{u}_n$, and the integration is over all values of $\mathbf{u}_1, \mathbf{u}_2, \ldots, \mathbf{u}_n$. This is the type of average that is being referred to, in general, when the overbar symbol is used in the sequel.

When the turbulence is spatially homogeneous, $\mathbf{u}$ is said to be a *stationary* random function of $\mathbf{x}$. The function P is then independent of the location (but not of the relative configuration) in space of the set of n points $(\mathbf{x}_1, t_1)$, $(\mathbf{x}_2, t_2)$, ..., $(\mathbf{x}_n, t_n)$; that is, the joint-probability distribution of the values of the velocity at the n points $(\mathbf{x}_1 + \mathbf{y}, t_1)$, $(\mathbf{x}_2 + \mathbf{y}, t_2)$, ..., $(\mathbf{x}_n + \mathbf{y}, t_n)$ is independent of the space vector $\mathbf{y}$. Likewise the average $\bar{F}$ is independent of the location in space of the n points.

The property of spatial homogeneity says, in effect, that all regions of space are similar so far as the statistical properties of $\mathbf{u}$ are concerned, and this suggests that the result of averaging over a large number of realizations or trials could be obtained equally well by averaging over a large region of space for *one* realization. Such a space average of the quantity F might be defined as

$$\lim_{V \to \infty} \frac{1}{V} \int F(\mathbf{u}_1, \mathbf{u}_2, \ldots, \mathbf{u}_n)\, d\mathbf{y} \qquad (2.1.2)$$

where $\mathbf{u}_1, \mathbf{u}_2, \ldots, \mathbf{u}_n$ are now the values of the velocity at the points $(\mathbf{x}_1 + \mathbf{y}, t_1)$, $(\mathbf{x}_2 + \mathbf{y}, t_2)$, ..., $(\mathbf{x}_n + \mathbf{y}, t_n)$ for any one realization of the field of turbulence, and V is the volume of $\mathbf{y}$-space over which the integral is taken. For this method of averaging to have any significance, it is necessary that the limit exist and that it be independent of the definition of the volume V. It is one of the objectives of ergodic theory to show that these conditions are in fact satisfied, and that the space average thereby obtained is the same for almost all† realizations of the field and is identical with the probability average (2.1.1), provided that the function F satisfies certain simple conditions (viz. provided F is finite and continuous in mean-square). Ergodic theory‡ is beyond the scope of this book and we shall be content to assume the result stated.

Consequently, for our problem of homogeneous turbulence and for suitable choices of the function F, the space average of F for almost any realization of the turbulence field is to be regarded as identical with the probability average, and either method of averaging may be employed at will. On the other hand, time averages play no part in our problem, although they will be relevant in the many problems of turbulent motion which is statistically steady.

These rules for taking averages apply to the idealized mathematical problem of homogeneous turbulence, and we must now consider whether they correspond to the procedure actually used in wind-tunnel experiments. As has already been mentioned, the usual method of generating turbulence in the wind tunnel is to place a grid, or regular array of bars, across the wind-tunnel stream. The resulting turbulence is statistically steady and is not spatially homogeneous, although it may be assumed homogeneous to a good approximation for most purposes. The decay with time in the idealized problem is thus replaced by decay with distance from the grid in the experimental field; stages of decay in the two problems may be compared by comparing the values of some simple average quantity, such as the mean kinetic energy. Measurements are made by placing a hot-wire anemometer at a point fixed relative to the grid and by recording the variation with time of the fluid velocity at the position of the hot wire. An average value of the required function of the velocity is then obtained by taking a mean over a long time with the aid of an appropriate instrument. Thus this experimental average is a time average for a field of turbulence

† That is, for all except a subset of realizations of relative measure zero.

‡ See A. Khintchine, 'Korrelationstheorie der stationären stochastischen Prozesse', *Math. Ann.* **109**, 1933, 604–15, and E. Hopf, 'Ergodentheorie', *Ergebn. Math.* Berlin, **5**, 1937, 63–151.

which is approximately but not exactly identical with our idealized field. Since the experimental field of turbulence is such that the velocity at a point fixed relative to the grid is a stationary random function of time, we can anticipate, again assuming the applicability of ergodic theory under suitable conditions, that a time average is identical with a probability average for the experimental field.†

It seems, then, that the conventional measuring methods provide a time average, and thereby a probability average, for the experimental, statistically steady field of turbulence, while our theory is concerned with a probability average for the idealized spatially homogeneous field (at the same stage of decay). Hence to the extent that the two fields of turbulence are statistically identical at corresponding stages of the decay, experimental averages will (almost always) be identical with the theoretical averages to be discussed in this book.

2.2. The complete statistical specification of the field of turbulence

We return now to the idealized field which is spatially homogeneous, and consider how this field may be specified. Granted that we know how averages should be taken, what mean quantities are required for a specification that determines the field statistically?

It is a premise of probability theory‡ that a random function $f(\alpha)$, say, defined for all values of α, is determined statistically by the

† There is a simple physical picture which is consistent with this conclusion, just as there is in the case of turbulence which is spatially homogeneous. If we inquire into the ultimate origin of the variation with time of the velocity in the turbulent flow downstream from the grid, we find that it must lie in the very small fluctuations in velocity which inevitably exist in the flow upstream of the grid. These very small fluctuations, which are not included in what we have called the ostensible data of the problem, act as a trigger for the unstable, steady flow that is produced in the wake of the grid. Thus with the passage of time there is a continuous variation of the boundary conditions (at the grid) of the field of turbulence generated on the downstream side of the grid. Moreover, this variation of the boundary conditions is random (the small velocity fluctuations on the upstream side of the grid are in fact a turbulent motion created by the various obstacles and irregularities in the wind tunnel). Hence the experimental time average is effectively a kind of probability average. However, we cannot be quite certain that a particular region of probability phase space is not being selected by the temporal variation of the boundary conditions at the grid, and so an appeal to the ergodic hypothesis is still necessary.

‡ See A. N. Kolmogoroff, *Grundbegriffe der Wahrscheinlichkeitsrechnung*, Berlin, 1933, and H. Cramér, *Mathematical methods of statistics*, Princeton, 1946.

complete system of joint-probability distributions of the values of the function at any n values of α, where n may take any integral value. Likewise the infinite field of turbulent motion is determined statistically by the complete system of joint-probability distributions of the values of the vector velocity $\mathbf{u}(\mathbf{x}, t)$ at any n points of space-time. If we require that the history of each realization of the velocity field be governed by the equations of momentum and continuity—as has already been done in § 1.2—this specification is clearly over-sufficient (although it may serve as a *description* of the motion). By formulating the dynamical problem as in § 1.2 we have in fact supposed that the statistical properties of the turbulence field at different values of t are uniquely related by the equations of momentum and continuity. Hence we shall usually be interested in the specification of the field of turbulence at a definite instant of time, and this will require a knowledge of the complete system of joint-probability distributions of the values of the velocity $\mathbf{u}$ at any n points of *space* at the appropriate value of t. Since the turbulence is spatially homogeneous these probability distributions depend only on the relative positions of the n points in space, and are independent of the location in space of the configuration of n points.

With this knowledge of how the field of turbulence may be specified statistically at any given value of t, we now know the maximum amount of data about the velocity field at the initial instant that must be supplied (in principle) in order that the mathematical problem of homogeneous turbulence as formulated in § 1.2 should be determinate. Likewise we know what information about the field at a variable value of t must be derived before the problem can be said to be completely solved. It is very probable that various physical conditions restrict the very wide class of turbulent motions implied by the above specification of a random function. For instance, the condition that the velocity is a continuous function of position and the incompressibility condition (1.2.1) each reduce the amount of data required to specify the field of turbulence. Moreover, we hope—and there is experimental evidence to support the hope—that the action of the Navier-Stokes equation of motion is to direct the random velocity field into a certain simple statistical state and so to restrict further the data needed to specify the ultimate velocity field.

Nevertheless, it is clear that we have little hope of being able to go far with the problem analytically unless we confine attention to the simplest kinds of averaged quantity (i.e. those corresponding to small values of n). The latter plan has usually been adopted in past research on the grounds of expediency, but we should recognize its limitations. Unless the field of turbulence is specified *completely* at some initial instant, or unless it should turn out that there are some features of the turbulence that are dynamically independent of each other, it is impossible to get a determinate set of dynamical equations for any of the quantities describing the turbulence at any later instant. In other words we may expect to be unable to find sufficient dynamical equations to solve for the simple mean values under discussion.

2.3. Mean values of velocity products

It is readily seen that the joint-probability distribution of the values of the velocity at any n points is uniquely related to the complete set of *averaged products* of the values of the velocity at these n points, provided the probability distribution satisfies certain weak conditions. For example, for the simple case $n = 1$, we can define a characteristic function $\phi(\boldsymbol{\alpha})$ as the (three-dimensional) Fourier transform of the probability density function $P(\mathbf{u})$, i.e.

$$\phi(\boldsymbol{\alpha}) = \int e^{i\boldsymbol{\alpha}.\mathbf{u}} P(\mathbf{u})\, d\mathbf{u}$$

$$= \sum_{p=0}^{\infty} \frac{i^p}{p!} \overline{(\boldsymbol{\alpha}.\mathbf{u})^p},$$

provided it is possible to expand $\phi(\boldsymbol{\alpha})$ as a Taylor series in the components of $\boldsymbol{\alpha}$. The coefficients in the series are proportional to the mean value of products of the components of $\mathbf{u}$, and the complete set of coefficients in the series—which determines $P(\mathbf{u})$, provided the Fourier transformation can be inverted—is determined by the complete set of product mean values. Similar remarks can be made for an arbitrary value of n, with the aid of a $3n$-dimensional Fourier transformation of the joint-probability density function.

The averaged products of the values of the velocity at different points of space† are the principal working tools of the analysis in

† And, occasionally, at different times. The kinematical analysis in this and the next chapter is concerned with space intervals only, since the extensions necessary to allow also for time intervals are usually obvious.

succeeding chapters. The mean value of the product of m components of the velocities at n different points ($m \geqslant n$) will be called an 'm-order n-point product mean value'. These mean values are functions of the configuration formed by the n points (but not of the location of the configuration) and of the time of decay. When $n=1$ the accepted term for the mean value is 'velocity moment' (of the m-order). When $m=n=2$, the mean value becomes the covariance of the velocities at the two points, although the term 'velocity correlation' is very firmly established in the literature and will be used in this work. Less justifiably, the term 'triple-velocity correlation' is used in the literature for the product mean value given by $m=3$, $n=2$; there seems to be no risk of confusion in this term, so that in view of the special place of this mean value in the theory it might as well be retained.

Each of the m velocity components occurring in an m-order product mean value may be any one of three orthogonal components, and the general mean value has 3^m scalar components. Each of the m velocity vectors from which the product is formed transforms, under change of the coordinate system, like a tensor of the first order (and this property is retained after the linear operation of taking a mean has been made), so that the 3^m components of the m-order mean value form (in Euclidean space) a tensor of order m. We shall make use of this fact in notation and in the derivation of the general functional form of the mean value.

The configuration formed by the n points at which the velocities are taken can be specified by $n-1$ space vectors; for example, by the position of $n-1$ of the points relative to the remaining point. Hence with a Cartesian coordinate system and the usual tensor index notation, we can write the m-order n-point velocity-product mean value as

$$\overline{u_i(\mathbf{x}_1, t)\, u_j(\mathbf{x}_2, t) \dots u_p(\mathbf{x}_m, t)} = Q^{(m)}_{ij\dots p}(\mathbf{r}, t), \qquad (2.3.1)$$

where $Q^{(m)}_{ij\dots p}$ is a tensor of order m, and $\mathbf{r}$ is a $3(n-1)$-dimensional vector which specifies the configuration formed by those n of the points $\mathbf{x}_1, \mathbf{x}_2, \dots, \mathbf{x}_m$ that are different.

It will be seen later that it is also useful to consider the Fourier transforms (in the complex form, for neatness) of velocity-product mean values with respect to the space vectors on which they depend.

After taking $3(n-1)$ repeated scalar Fourier transforms of $Q^{(m)}_{ij\ldots p}$ (assuming for the moment that this operation is permissible) we have a (complex) quantity $\chi^{(m)}_{ij\ldots p}$ which is a function of the new $3(n-1)$-dimensional vector $\boldsymbol{\kappa}$:

$$\chi^{(m)}_{ij\ldots p}(\boldsymbol{\kappa}, t) = \left(\frac{1}{2\pi}\right)^{3(n-1)} \int Q^{(m)}_{ij\ldots p}(\mathbf{r}, t)\, e^{-\iota \boldsymbol{\kappa}.\mathbf{r}}\, d\mathbf{r}, \qquad (2.3.2)$$

where $\iota = \sqrt{-1}$, $d\mathbf{r}$ represents an element of volume in $\mathbf{r}$-space and the integration is over all values of $\mathbf{r}$. $\chi^{(m)}_{ij\ldots p}$ is a linear functional of $Q^{(m)}_{ij\ldots p}$ and hence is also a tensor of order m. $\chi^{(m)}_{ij\ldots p}$ and $Q^{(m)}_{ij\ldots p}$ are also related by the Fourier transform relation inverse to (2.3.2), viz.

$$Q^{(m)}_{ij\ldots p}(\mathbf{r}, t) = \int \chi^{(m)}_{ij\ldots p}(\boldsymbol{\kappa}, t)\, e^{\iota \boldsymbol{\kappa}.\mathbf{r}}\, d\boldsymbol{\kappa}. \qquad (2.3.3)$$

The unsymmetrical distribution of 2π factors in (2.3.2) and (2.3.3) is chosen (consistently) for later convenience.

The quantity $\chi^{(m)}_{ij\ldots p}$ will exist when the integral $\int |Q^{(m)}_{ij\ldots p}|\, d\mathbf{r}$ converges.† This will not be the case, in general, when $m > 3$ (nor when $m \leqslant 3$, if the mean velocity is not zero everywhere), for then the velocity product can be split into two parts, each of which has a non-zero mean irrespective of the distance between the two sub-configurations on which the two parts depend. In such cases, we must first subtract from the mean value (2.3.1) various products of lower order mean values formed from

$$u_i(\mathbf{x}_1),\ u_j(\mathbf{x}_2),\ \ldots\ u_p(\mathbf{x}_m),$$

chosen in such a way that the resulting quantity vanishes whenever $\mathbf{r}^2 \to \infty$ (which procedure is equivalent to considering the m-order *cumulant* tensor of the velocities $\mathbf{u}(\mathbf{x}_1)$, $\mathbf{u}(\mathbf{x}_2)$, … $\mathbf{u}(\mathbf{x}_n)$). The condition for the Fourier transform to exist will normally be satisfied by this reduced velocity product mean value, or cumulant, and we shall regard the symbol $Q^{(m)}_{ij\ldots p}$ as applying to the cumulant whenever the question of taking the Fourier transform of velocity product mean values is involved.

Almost all the existing analysis of homogeneous turbulence can be carried out either in terms of velocity-product mean values or

† See E. C. Titchmarsh, *Introduction to the theory of Fourier integrals*, Oxford University Press, 1937.

in terms of their Fourier transforms. The choice of function to be used will usually depend on the type of equation under discussion, and there will sometimes be a considerable gain in simplicity of the analysis with one particular kind of function. However, we should not lose sight of the fact that in general it is the product mean values that are measured in the laboratory.

In future work we shall also need to consider the mean value of the product of velocities and spatial derivatives of velocities at different points. A simple example of such a mean value is

$$\overline{u_i(\mathbf{x}_1, t)\frac{\partial u_j(\mathbf{x}_2, t)}{\partial(\mathbf{x}_2)_q} \ldots u_p(\mathbf{x}_m, t)}.$$

Provided none of the factors in the mean value, other than $u_j(\mathbf{x}_2, t)$, depends on $\mathbf{x}_2$ (that is, provided $\mathbf{x}_2$ is different from $\mathbf{x}_1, \mathbf{x}_3, \ldots, \mathbf{x}_m$), and since the operations of averaging and differentiating may readily be shown to permute, this mean value is equal to

$$\frac{\partial}{\partial(\mathbf{x}_2)_q} Q^{(m)}_{ij\ldots p}(\mathbf{r}, t),$$

and further reduction depends on the manner in which $\mathbf{r}$ depends on $\mathbf{x}_2$. If the $n-1$ subvectors which comprise the $3(n-1)$-dimensional vector $\mathbf{r}$ are the positions of $n-1$ of the points of the configuration relative to $\mathbf{x}_1$, we have

$$\frac{\partial}{\partial(\mathbf{x}_2)_q} = \frac{\partial}{\partial(\mathbf{r}_1)_q}, \quad \frac{\partial}{\partial(\mathbf{x}_1)_q} = -\left[\frac{\partial}{\partial(\mathbf{r}_1)_q} + \ldots + \frac{\partial}{\partial(\mathbf{r}_{n-1})_q}\right]$$

for operations on mean values such as $Q^{(m)}_{ij\ldots p}(\mathbf{r}, t)$.

The spatial derivative of a scalar quantity transforms, under a change of coordinate system, like a tensor of the first order. Likewise the spatial derivative of a tensor of order m is itself a tensor of order $m+1$, and each extra differentiation adds one to the order of the resulting tensor quantity. Thus $\frac{\partial}{\partial(\mathbf{r}_1)_q} Q^{(m)}_{ij\ldots p}$ is a tensor of order $m+1$, and (2.3.3) shows that its Fourier transform is the tensor $\iota(\boldsymbol{\kappa}_1)_q \chi^{(m)}_{ij\ldots p}$, also of order $m+1$ (where $\boldsymbol{\kappa}_1, \ldots, \boldsymbol{\kappa}_{n-1}$ are the vectors complementary to $\mathbf{r}_1, \ldots, \mathbf{r}_{n-1}$ in the Fourier transformation). We see that in general differentiation of a velocity-product mean value with respect to one of the components of $\mathbf{r}$ is equivalent to multiplication of its Fourier transform by (ι times) the corresponding component of $\boldsymbol{\kappa}$; this elimination of differentials is the reason why analysis in terms of Fourier transforms is frequently simpler.

An example of mean values involving velocity derivatives is provided by the introduction of the incompressibility condition (1.2.1), viz.

$$\frac{\partial u_i(\mathbf{x}, t)}{\partial x_i} = 0,$$

where the repeated index i is to be regarded as being summed over all three possible values of the index.† One of the conditions which the tensor $Q^{(m)}_{ij\ldots p}(\mathbf{r}, t)$ must satisfy as a consequence is

$$\frac{\partial}{\partial(\mathbf{x}_2)_j} Q^{(m)}_{ij\ldots p}(\mathbf{r}, t) = \frac{\partial}{\partial(\mathbf{r}_1)_j} Q^{(m)}_{ij\ldots p}(\mathbf{r}, t) = 0, \qquad (2.3.4)$$

and there is a similar equation for each of the vectors (except $\mathbf{x}_1$) which occur singly in the group $\mathbf{x}_1, \mathbf{x}_2, \ldots, \mathbf{x}_m$. If $\mathbf{x}_1$ occurs singly, we have

$$\frac{\partial}{\partial(\mathbf{x}_1)_i} Q^{(m)}_{ij\ldots p}(\mathbf{r}, t) = -\left[\frac{\partial}{\partial(\mathbf{r}_1)_i} + \ldots + \frac{\partial}{\partial(\mathbf{r}_{n-1})_i}\right] Q^{(m)}_{ij\ldots p}(\mathbf{r}, t) = 0. \qquad (2.3.5)$$

The conditions on $\chi^{(m)}_{ij\ldots p}(\boldsymbol{\kappa}, t)$ corresponding to (2.3.4) and (2.3.5) are

$$(\boldsymbol{\kappa}_1)_j \chi^{(m)}_{ij\ldots p}(\boldsymbol{\kappa}, t) = 0, \qquad (2.3.6)$$

and

$$[(\boldsymbol{\kappa}_1)_i + \ldots + (\boldsymbol{\kappa}_{n-1})_i] \chi^{(m)}_{ij\ldots p}(\boldsymbol{\kappa}, t) = 0; \qquad (2.3.7)$$

the consequent restrictions on the possible form of $\chi^{(m)}_{ij\ldots p}(\boldsymbol{\kappa}, t)$ are readily taken into account, as will be seen later.

2.4. General properties of the velocity correlation and spectrum tensors

The velocity correlation tensor (i.e. the second-order two-point product mean value) and its Fourier transform are of special importance in turbulence theory. The velocity correlation is the simplest of the product mean values, apart from the almost trivial one-point product means, and its physical significance (and, even more so, that of its Fourier transform) is easily understood.

† This convenient summation convention will be used without explanation in future.

Furthermore, we shall see later that the joint-probability distribution of the velocities at two points is not very different from a normal distribution, so that a knowledge of the velocity correlation carries with it an approximate knowledge of the higher order mean values involving the velocities at two points. In this section we shall consider some of the general mathematical results concerning the velocity correlation, as a preliminary to several of the later chapters in which the analysis will be carried further making use of the particular properties of turbulence.

The velocity correlation tensor for two points separated by the space vector $\mathbf{r}$ (which is the three-dimensional vector defining the configuration of two points) is

$$R_{ij}(\mathbf{r}) = \overline{u_i(\mathbf{x})\, u_j(\mathbf{x}+\mathbf{r})};$$

the dependence on t will not be shown explicitly unless different instants are under consideration. An immediate geometrical property of the velocity correlation in homogeneous turbulence is

$$R_{ij}(\mathbf{r}) = R_{ji}(-\mathbf{r}). \tag{2.4.1}$$

We shall assume (on the grounds, already stated in Chapter 1, that discontinuities could not continue to exist in the presence of viscous forces) that $\mathbf{u}(\mathbf{x})$ is a *continuous* function of $\mathbf{x}$, from which it follows that

$$\lim_{\mathbf{h}\to 0} \overline{u_i(\mathbf{x})\,[u_j(\mathbf{x}+\mathbf{r}+\mathbf{h}) - u_j(\mathbf{x}+\mathbf{r})]} = 0,$$

i.e.

$$\lim_{\mathbf{h}\to 0} [R_{ij}(\mathbf{r}+\mathbf{h}) - R_{ij}(\mathbf{r})] = 0,$$

showing that $R_{ij}(\mathbf{r})$ is continuous† at all values of $\mathbf{r}$. According to the Schwarzian inequality,

$$R_{ij}(\mathbf{r}) \leqslant [\overline{u_i^2(\mathbf{x})}\,.\,\overline{u_j^2(\mathbf{x}+\mathbf{r})}]^{\frac{1}{2}}$$

$$= [R_{ii}(0)\,.\,R_{jj}(0)]^{\frac{1}{2}} \quad \text{(no summation convention here)},$$

and for the particular case $i=j$, we have

$$R_{ii}(\mathbf{r}) \leqslant R_{ii}(0).$$

The properties of a general stationary random function (more particularly, of a scalar function of a scalar argument) have been

† The same result can be established under conditions less restrictive—but also less relevant—than the assumption that $\mathbf{u}(\mathbf{x})$ is continuous.

studied intensively by mathematicians interested in probability theory, and it is not proposed to refer to all the results† here. However, one fundamental theorem concerning the (velocity) correlation, established by Cramér,‡ must be quoted. This theorem is an extension, to the case of a multi-dimensional random process, of a well-known result due to Khintchine,§ and may be stated as follows:

'The necessary and sufficient condition that $R_{ij}(\mathbf{r})$ should be the correlation tensor of a continuous|| stationary random process is that it should be expressible in the form

$$R_{ij}(\mathbf{r}) = \int \Phi_{ij}(\boldsymbol{\kappa})\, e^{i\boldsymbol{\kappa}.\mathbf{r}}\, d\boldsymbol{\kappa}, \qquad (2.4.2)¶$$

where $\Phi_{ij}(\boldsymbol{\kappa})$ is a complex tensor such that

$$(a) \quad \int |\Phi_{ij}(\boldsymbol{\kappa})|\, d\boldsymbol{\kappa} < \infty,$$

and $$(b) \quad \Phi = X_i X_j^* \Phi_{ij}(\boldsymbol{\kappa})$$

is a non-negative quadratic form (i.e. $\Phi \geqslant 0$ for an arbitrary choice of the complex constants X_i).'

In this statement, $d\boldsymbol{\kappa}$ is written for $d\kappa_1 d\kappa_2 d\kappa_3$, the integrals are taken over the whole of wave-number space, and X_i^* denotes the complex conjugate of X_i.

† Basic theorems will be found in: P. Lévy, *Processus stochastique et mouvements brownien*, vol. 1, Paris, 1948, and K. Karhunen, 'Uber lineare Methoden in der Wahrscheinlichkeitsrechnung', *Ann. Acad. Sci. fenn.* A, no. 37, 1947. Reviews written from a less purely mathematical point of view are: J. Kampé de Fériet (1939) and J. E. Moyal, 'Stochastic processes and statistical physics', *J. R. Statist. Soc.* Series B, **11**, 1949, 150–210.

‡ H. Cramér, 'On the theory of stationary random processes', *Ann. Math.* **41**, 1940, 215–30.

§ Loc. cit. (p. 16).

|| That is, relating to a function $\mathbf{u}$ which is defined over a continuous range of values of the argument $\mathbf{x}$.

¶ A more general form of the condition (2.4.2) is that $R_{ij}(\mathbf{r})$ should be expressible as $\int e^{i\boldsymbol{\kappa}.\mathbf{r}}\, d\Psi_{ij}(\boldsymbol{\kappa})$, where the function $\Psi_{ij}(\boldsymbol{\kappa})$ is not necessarily differentiable. However, we shall see from the interpretation of $\Phi_{ij}(\boldsymbol{\kappa})$ as a density of kinetic energy in wave-number space that $\Phi_{ij}(\boldsymbol{\kappa})$ can safely be assumed in the present context to be a continuous function of $\boldsymbol{\kappa}$. The basis for this assumption (which restricts the analysis essentially in notation only) is that the non-linear dynamical processes would immediately spread the energy of a line or step in the spectrum over a continuous range of wave-numbers. See also § 5.2.

For the purposes of our problem we begin with the premise that $R_{ij}(\mathbf{r})$ *is* a correlation tensor describing homogeneous turbulence, so that Cramér's theorem establishes that there exists a function $\Phi_{ij}(\boldsymbol{\kappa})$ with the properties (*a*) and (*b*). In view of (*a*) we can write

$$\Phi_{ij}(\boldsymbol{\kappa}) = \frac{1}{8\pi^3}\int R_{ij}(\mathbf{r})\, e^{-i\boldsymbol{\kappa}.\mathbf{r}}\, d\mathbf{r} \tag{2.4.3}$$

(as mentioned in §2.3 the existence of the Fourier transform of a product mean value usually needs to be assumed, but in this case of a second-order product it is established rigorously), and then (2.4.1) shows that

$$\Phi_{ij}(\boldsymbol{\kappa}) = \Phi_{ji}(-\boldsymbol{\kappa}) = \Phi^*_{ji}(\boldsymbol{\kappa}), \tag{2.4.4}$$

i.e. $\Phi_{ij}(\boldsymbol{\kappa})$ is a tensor with Hermitian symmetry and

$$\Phi = X_i X_j^* \Phi_{ij}(\boldsymbol{\kappa})$$

is an Hermitian form.

When $|\mathbf{r}| = 0$, equation (2.4.2) becomes

$$R_{ij}(0) = \overline{u_i(\mathbf{x})\, u_j(\mathbf{x})} = \int \Phi_{ij}(\boldsymbol{\kappa})\, d\boldsymbol{\kappa}, \tag{2.4.5}$$

showing that $\Phi_{ij}(\boldsymbol{\kappa})$ represents a density, in wave-number space, of contributions to $\overline{u_i(\mathbf{x})\, u_j(\mathbf{x})}$. A knowledge of all components of the tensor $\overline{u_i(\mathbf{x})\, u_j(\mathbf{x})}$ is necessary and sufficient to determine the energy (per unit mass of fluid) associated with an arbitrary component of the velocity, and we shall therefore call it the *energy tensor*. Thus $\Phi_{ij}(\boldsymbol{\kappa})$ describes a distribution of energy in $\boldsymbol{\kappa}$-space. Its physical significance is not given unambiguously by (2.4.5), but we shall see in the next section that it describes how the energy associated with each velocity component is distributed over the various wave-numbers in a harmonic resolution of the velocity field, and $\Phi_{ij}(\boldsymbol{\kappa})$ will be called the *energy spectrum tensor*. It is this physical significance that makes $\Phi_{ij}(\boldsymbol{\kappa})$—or its Fourier transform $R_{ij}(\mathbf{r})$—the most important single quantity describing the field of turbulence.

For an incompressible fluid, (1.2.1) shows that

$$\overline{u_i(\mathbf{x})\frac{\partial u_j(\mathbf{x}+\mathbf{r})}{\partial r_j}} = 0.$$

From this and a similar relation we find the two continuity conditions

$$\frac{\partial R_{ij}(\mathbf{r})}{\partial r_j} = \frac{\partial R_{ij}(\mathbf{r})}{\partial r_i} = 0. \tag{2.4.6}$$

The corresponding conditions for the spectrum tensor are found from (2.4.2) to be

$$\kappa_j \Phi_{ij}(\boldsymbol{\kappa}) = \kappa_i \Phi_{ij}(\boldsymbol{\kappa}) = 0. \tag{2.4.7}$$

In view of the Hermitian symmetry (2.4.4), only one of these continuity conditions on $R_{ij}(\mathbf{r})$ or $\Phi_{ij}(\boldsymbol{\kappa})$ is independent.

J. Kampé de Fériet (1948) has shown that the definition of the spectrum tensor and the continuity condition (2.4.7) together impose upon $\Phi_{ij}(\boldsymbol{\kappa})$ a certain functional form. It is known from the theory of quadratic forms that it is always possible to find a transformation defined by

$$Y_1 = a'_i X_i, \quad Y_2 = b'_i X_i, \quad Y_3 = c'_i X_i,$$

where the unit complex vectors $\mathbf{a}'$, $\mathbf{b}'$ and $\mathbf{c}'$ are orthogonal, i.e.

$$a'_i b_i^{*\prime} = b'_i c_i^{*\prime} = c'_i a_i^{*\prime} = 0$$

the transformation being such that the quadratic form Φ is reduced to its diagonal form

$$\Phi = X_i X_j^* \Phi_{ij}(\boldsymbol{\kappa}) = s_1 Y_1 Y_1^* + s_2 Y_2 Y_2^* + s_3 Y_3 Y_3^*, \tag{2.4.8}$$

where s_1, s_2 and s_3 are real functions of $\boldsymbol{\kappa}$. (When $\Phi_{ij}(\boldsymbol{\kappa})$ and X_i are real, this operation amounts merely to finding the direction cosines a'_i, b'_i, c'_i of the principal axes of the quadratic surface $\Phi = \text{constant}$.) Cramér's theorem requires that Φ be non-negative for all X_i, and consequently for all Y_i, so that each of the coefficients s_1, s_2 and s_3 must be non-negative. It appears that the meaning of property (*b*) in Cramér's theorem is that when referred to principal axes, and therefore always, the diagonal elements of $\Phi_{ij}(\boldsymbol{\kappa})$ are separately non-negative—as indeed we should expect if the diagonal elements represent a density in wave-number space of contributions to kinetic energy.

The transformed expression (2.4.8) is valid for arbitrary X_i, and if we choose the particular case in which the vector $\mathbf{X}$ is parallel to $\boldsymbol{\kappa}$ the condition of incompressibility (2.4.7) shows that Φ is then zero. Since s_1, s_2 and s_3 are each non-negative, a zero value of Φ requires each term on the right side of (2.4.8) to be zero. Not more than two of the orthogonal components Y_1, Y_2, Y_3 can vanish (since $Y_i Y_i^* = X_i X_i^*$), so that we require†

$$Y_1 = Y_2 = 0, \quad s_3 = 0,$$

or, since **X** is parallel to **κ**,

$$a_i' \kappa_i = b_i' \kappa_i = 0, \quad s_3 = 0.$$

Hence the expression for Φ becomes

$$\Phi = X_i X_j^* \Phi_{ij}(\boldsymbol{\kappa}) = s_1 Y_1 Y_1^* + s_2 Y_2 Y_2^*,$$

showing that the general form of $\Phi_{ij}(\boldsymbol{\kappa})$ is

$$\Phi_{ij}(\boldsymbol{\kappa}) = a_i a_j^* + b_i b_j^*, \tag{2.4.9}$$

where **κ**, **a** $(=\sqrt{(s_1)}\,\mathbf{a}')$ and **b** $(=\sqrt{(s_2)}\,\mathbf{b}')$ are orthogonal (complex) vectors, **a** and **b** being functions of **κ**.

An alternative version of (2.4.9) is obtained by making use of the identity‡

$$\frac{\kappa_i \kappa_j}{\boldsymbol{\kappa}^2} + \frac{a_i a_j^*}{\mathbf{a}^2} + \frac{b_i b_j^*}{\mathbf{b}^2} = \delta_{ij}$$

to eliminate $b_i b_j^*$, viz.

$$\Phi_{ij}(\boldsymbol{\kappa}) = \mathbf{b}^2 \left(\delta_{ij} - \frac{\kappa_i \kappa_j}{\boldsymbol{\kappa}^2}\right) + a_i a_j^* \left(1 - \frac{\mathbf{b}^2}{\mathbf{a}^2}\right). \tag{2.4.10}$$

When the turbulence is isotropic, no direction in the plane normal to **κ** can be preferred, so that $\mathbf{a}^2 = \mathbf{b}^2$, leaving only the first term of (2.4.10) in this case, as will be found by other methods in Chapter III.

2.5. Fourier analysis of the velocity field

If we are to make use of the ideas of statistical mechanics, and if we are to make hypotheses involving the notion of statistical equilibrium, it is necessary first to decide what we mean by a component, or degree of freedom, of the turbulent motion. We should like to find some method of resolving the motion into a number of components which, for preference, make additive

† For real quantities the corresponding geometrical interpretation is that Φ = constant represents an elliptic cylinder with its generators parallel to **κ**.

‡ δ_{ij} is the unit diagonal tensor, which takes the values

$$\delta_{ij} = 1 \text{ when } i = j,$$

and $$\delta_{ij} = 0 \text{ when } i \neq j.$$

contributions to the energy of the motion, and have a clearly recognizable physical meaning. The demand for additive contributions to the energy suggests immediately that we should represent the velocity distribution by means of a set of orthogonal functions; trigonometric functions are then the obvious choice,† since they neither decrease nor increase their amplitude at infinity, like the function which we wish to represent.

Fourier analysis of the velocity field provides us with an extremely valuable analytical tool and one which is well-nigh indispensable for the interpretation of equilibrium or similarity hypotheses. Since the wave-length is the parameter specifying the different Fourier components, Fourier analysis corresponds, in a general way, to a resolution into components of the motion of different linear *size*. It also gives a definite meaning to the idea of the different degrees of freedom possessed by the fluid. Large-scale and small-scale components of the motion are not attached to limited portions of the fluid in the way that different degrees of freedom of a simple gas are attached to different molecules, but nevertheless we can think of the turbulent motion as consisting of the superposition of a large number of different-sized component motions, which make additive contributions to the total energy and which interact with each other in a way demanded by the non-linear term in the equation of motion. The precise form of motion associated with each component is not of great importance from a physical point of view, but it will be seen to be a sinusoidal shearing motion, or transverse plane 'wave' (which does not propagate in time, of course).

Consider first the Fourier analysis of one realization of the whole velocity field at a given value of t. If $\mathbf{u}(\mathbf{x})$ were periodic in $\mathbf{x}$, the appropriate representation would be as a Fourier series. If

$$\int |\mathbf{u}(\mathbf{x})| \, d\mathbf{x}$$

taken over the whole field were bounded, a Fourier integral would be appropriate. Neither of these conditions is satisfied when the

† It can be shown that the choice of trigonometric functions is in fact *necessary* if we wish to have a method of representation which is valid for any stationary random function.

velocity is a stationary random function of position,† and a modification of these representations is necessary. Consider the function $\mathbf{u}(\mathbf{x}, X)$, where X is a disposable parameter, defined by

$$\mathbf{u}(\mathbf{x}, X) = \mathbf{u}(\mathbf{x}) \quad \text{for} \quad -X \leqslant x_i \leqslant X,$$
$$\mathbf{u}(\mathbf{x}, X) = 0 \qquad \text{for} \quad |x_i| > X,$$

which satisfies the condition for a Fourier integral to exist. The corresponding (vector) Fourier coefficient for wave-number $\boldsymbol{\kappa}$ is

$$A(\boldsymbol{\kappa}, X) = \frac{1}{(2\pi)^3}\int \mathbf{u}(\mathbf{x}, X)\, e^{-\iota\boldsymbol{\kappa}.\mathbf{x}}\, d\mathbf{x},$$

the integration being over all $\mathbf{x}$-space. The difficulty to be faced in a Fourier analysis of the real velocity distribution is that $\mathbf{A}(\boldsymbol{\kappa}, X)$ diverges as $X \to \infty$. The device that is adopted depends on the fact that $\mathbf{A}(\boldsymbol{\kappa}, X)$ is a rapidly oscillating function of $\boldsymbol{\kappa}$ and that, as can be shown‡ rigorously, the limit of

$$\int_{\kappa_1'}^{\kappa_1''}\int_{\kappa_2'}^{\kappa_2''}\int_{\kappa_3'}^{\kappa_3''} \mathbf{A}(\boldsymbol{\kappa}, X)\, d\kappa_1\, d\kappa_2\, d\kappa_3$$

as $X \to \infty$ does exist; defining this limit as $[\mathbf{Z}(\boldsymbol{\kappa})]_{\boldsymbol{\kappa}'}^{\boldsymbol{\kappa}''}$, we have

$$[\mathbf{Z}(\boldsymbol{\kappa})]_{\boldsymbol{\kappa}'}^{\boldsymbol{\kappa}''} = \frac{1}{(2\pi)^3}\int \mathbf{u}(\mathbf{x})\left(\frac{e^{-\iota\kappa_1''x_1} - e^{-\iota\kappa_1'x_1}}{-\iota x_1}\right) \times \left(\frac{e^{-\iota\kappa_2''x_2} - e^{-\iota\kappa_2'x_2}}{-\iota x_2}\right)\left(\frac{e^{-\iota\kappa_3''x_3} - e^{-\iota\kappa_3'x_3}}{-\iota x_3}\right) d\mathbf{x}. \qquad (2.5.1)$$

(More precisely, the limit exists for almost every realization of the velocity field, viz. for those realizations for which space averages exist. It is necessary to regard (2.5.1) as a *stochastic* integral, i.e. the corresponding sum converges to a quantity $[\mathbf{Z}(\boldsymbol{\kappa})]_{\boldsymbol{\kappa}'}^{\boldsymbol{\kappa}''}$ in the sense that the mean square of the difference between $[\mathbf{Z}(\boldsymbol{\kappa})]_{\boldsymbol{\kappa}'}^{\boldsymbol{\kappa}''}$ and the sum tends to zero as the number of terms in the summation

† But both may be satisfied by considering hypothetical velocity fields which differ from the real field in physically trivial respects only; if $\mathbf{u}(\mathbf{x})$ is regarded as being periodic with very large wave-length a Fourier series may be used, and if $\mathbf{u}(\mathbf{x})$ is regarded as being zero outside a very large box a Fourier integral may be used. The only practical disadvantage of these two procedures—which have been widely used—is that the orders of magnitude of quantities depending on the large length which is introduced are not always evident.

‡ N. Wiener, 'Generalized harmonic analysis', *Acta Math.* **55**, 1930, 117–258.

approaches infinity.)† When $\boldsymbol{\kappa}'$ and $\boldsymbol{\kappa}''$ differ by a small quantity $d\boldsymbol{\kappa}$, we write

$$d\mathbf{Z}(\boldsymbol{\kappa}) = [\mathbf{Z}(\boldsymbol{\kappa})]_{\boldsymbol{\kappa}}^{\boldsymbol{\kappa}+d\boldsymbol{\kappa}}$$
$$= \frac{1}{(2\pi)^3}\int \mathbf{u}(\mathbf{x})\, e^{-\iota\boldsymbol{\kappa}.\mathbf{x}} \left(\frac{e^{-\iota d\kappa_1 x_1}-1}{-\iota x_1}\right)\left(\frac{e^{-\iota d\kappa_2 x_2}-1}{-\iota x_2}\right)\left(\frac{e^{-\iota d\kappa_3 x_3}-1}{-\iota x_3}\right) d\mathbf{x} \tag{2.5.2}$$

(so that if the derivative $\mathbf{A}(\boldsymbol{\kappa})$ existed, we should have

$$d\mathbf{Z}(\boldsymbol{\kappa}) = \mathbf{A}(\boldsymbol{\kappa})\, d\kappa_1\, d\kappa_2\, d\kappa_3$$

to the first order in $d\boldsymbol{\kappa}$).

The relation inverse to (2.5.2) is

$$\mathbf{u}(\mathbf{x}) = \int e^{\iota\boldsymbol{\kappa}.\mathbf{x}}\, d\mathbf{Z}(\boldsymbol{\kappa}), \tag{2.5.3}$$

where the integration is over all wave-number space. The ordinary conception of an integral is not adequate here, since the derivative of the function $\mathbf{Z}(\boldsymbol{\kappa})$ is not finite; (2.5.3) must be read as a (stochastic) Fourier-Stieltjes integral, of a generalized kind‡ to take account of the fact that when the energy spectrum is continuous the function $\mathbf{Z}(\boldsymbol{\kappa})$ is not in general of bounded total variation. For the purposes of the present brief account of harmonic analysis it will be sufficient to regard (2.5.3) as a formal consequence of (2.5.2); when we come to consider the average properties of the function $d\mathbf{Z}(\boldsymbol{\kappa})$ we shall find that the difficulties associated with its peculiar behaviour disappear.

The increment $d\mathbf{Z}(\boldsymbol{\kappa})$ is a random variable, since its value at any $\boldsymbol{\kappa}$ depends on the particular realization of the velocity distribution $\mathbf{u}(\mathbf{x})$, and we are interested primarily in its average properties. The covariance of the components of $[\mathbf{Z}(\boldsymbol{\kappa})]_{\boldsymbol{\kappa}'}^{\boldsymbol{\kappa}''}$ is found from (2.5.1) to be

$$\overline{[Z_i^*(\boldsymbol{\kappa})]_{\boldsymbol{\kappa}'}^{\boldsymbol{\kappa}''}[Z_j(\boldsymbol{\kappa})]_{\boldsymbol{\kappa}'}^{\boldsymbol{\kappa}''}}$$
$$= \frac{1}{(2\pi)^6}\iint \overline{u_i(\mathbf{x})\,u_j(\mathbf{x}')} \prod_{n=1}^{3}\left(\frac{e^{\iota\kappa_n'' x_n}-e^{\iota\kappa_n' x_n}}{\iota x_n}\right)\left(\frac{e^{-\iota\kappa_n'' x_n'}-e^{-\iota\kappa_n' x_n'}}{-\iota x_n'}\right) d\mathbf{x}\, d\mathbf{x}'$$
$$= \frac{1}{(2\pi)^6}\iint R_{ij}(\mathbf{r}) \prod_{n=1}^{3}\left(\frac{e^{\iota\kappa_n'' x_n}-e^{\iota\kappa_n' x_n}}{\iota x_n}\right)\left(\frac{e^{-\iota\kappa_n''(x_n+r_n)}-e^{-\iota\kappa_n'(x_n+r_n)}}{-\iota(x_n+r_n)}\right) d\mathbf{x}\, d\mathbf{r}$$
$$= \frac{1}{(2\pi)^3}\int R_{ij}(\mathbf{r}) \prod_{n=1}^{3}\left(\frac{e^{-\iota\kappa_n' r_n}-e^{-\iota\kappa_n'' r_n}}{\iota r_n}\right) d\mathbf{r}. \tag{2.5.4}$$

† For an account of an approach to spectral analysis directly in terms of stochastic Fourier-Stieltjes integrals, see the Note by M. Loève on 'Fonctions aléatoires du second ordre' at the end of the book by P. Lévy (op. cit. on p. 25). An alternative approach which makes use of the ideas of Hilbert space is described by J. E. Moyal (loc. cit. on p. 25).

‡ See N. Wiener, loc. cit. (p. 30), and J. Kampé de Fériet (1949).

Putting $\boldsymbol{\kappa}'' - \boldsymbol{\kappa}' = d\boldsymbol{\kappa}$ and $\boldsymbol{\kappa}' = \boldsymbol{\kappa}$ as before, we find that both sides of (2.5.4) vanish with $d\boldsymbol{\kappa}$, but that

$$\lim_{d\boldsymbol{\kappa}\to 0} \frac{\overline{dZ_i^*(\boldsymbol{\kappa})\, dZ_j(\boldsymbol{\kappa})}}{d\kappa_1\, d\kappa_2\, d\kappa_3} = \frac{1}{(2\pi)^3}\int R_{ij}(\mathbf{r})\, e^{-i\boldsymbol{\kappa}.\mathbf{r}} d\mathbf{r},$$
$$= \Phi_{ij}(\boldsymbol{\kappa}), \qquad (2.5.5)$$

in view of the definition of $\Phi_{ij}(\boldsymbol{\kappa})$ in (2.4.3). A calculation similar to that leading to (2.5.4) shows that $\overline{dZ_i^*(\boldsymbol{\kappa}')\, dZ_j(\boldsymbol{\kappa}'')}$ is identically zero unless $\boldsymbol{\kappa}'$ and $\boldsymbol{\kappa}''$ are so nearly equal that the volume increments $d\boldsymbol{\kappa}'$ and $d\boldsymbol{\kappa}''$ overlap in wave-number space, showing that $d\mathbf{Z}(\boldsymbol{\kappa}')$ and $d\mathbf{Z}(\boldsymbol{\kappa}'')$ are statistically orthogonal.

The quantity $\Phi_{ij}(\boldsymbol{\kappa})$, introduced as the Fourier transform of the velocity correlation, is thus found to bear a close relation to the Fourier coefficients of the velocity distribution. $d\mathbf{Z}(\boldsymbol{\kappa})\, e^{i\boldsymbol{\kappa}.\mathbf{x}}$ is the contribution to the velocity field from an element of volume $d\boldsymbol{\kappa}$ in wave-number space, and since these contributions are statistically orthogonal the left side of (2.5.5) represents the density in wave-number space of contributions to the energy tensor $\overline{u_i(\mathbf{x})\, u_j(\mathbf{x})}$. This is the sense in which the tensor $\Phi_{ij}(\boldsymbol{\kappa})$ describes the spectrum of the kinetic energy of the turbulence. Analytically, we have

$$\overline{u_i(\mathbf{x})\, u_j(\mathbf{x})} = \iint e^{i\mathbf{x}.(\boldsymbol{\kappa}''-\boldsymbol{\kappa}')}\, \overline{dZ_i^*(\boldsymbol{\kappa}')\, dZ_j(\boldsymbol{\kappa}'')}$$
$$= \int \Phi_{ij}(\boldsymbol{\kappa})\, d\boldsymbol{\kappa}. \qquad (2.5.6)$$

For an incompressible fluid, we find from (2.5.3) that

$$\boldsymbol{\kappa}.\, d\mathbf{Z}(\boldsymbol{\kappa}) = 0. \qquad (2.5.7)$$

Hence the vector $d\mathbf{Z}(\boldsymbol{\kappa})$ can be expressed in terms of two orthogonal components each of which is orthogonal to $\boldsymbol{\kappa}$, and the link between (2.5.5) and the general form (2.4.9) for $\Phi_{ij}(\boldsymbol{\kappa})$ is immediately clear. If the fluid is not assumed to be incompressible, $d\mathbf{Z}(\boldsymbol{\kappa})$ has, in general, a component parallel to $\boldsymbol{\kappa}$, which represents a plane longitudinal or compression 'wave' (again we are speaking only of the instantaneous distribution of velocity), and which will make a contribution to $\Phi_{ij}(\boldsymbol{\kappa})$. The modifications to the above analysis which are necessary when the fluid is compressible have recently been considered by Moyal (1952).

We shall have occasion in later chapters to consider other properties of $d\mathbf{Z}(\boldsymbol{\kappa})$—in particular, the statistical connexion between values of $d\mathbf{Z}(\boldsymbol{\kappa})$ at different values of $\boldsymbol{\kappa}$.

The reader who is interested in a more complete mathematical discussion of the spectral analysis of stationary random functions should consult the works already cited, beginning with the paper by Wiener.

CHAPTER III

THE KINEMATICS OF HOMOGENEOUS TURBULENCE

Now that we have considered in a general way the mathematical representation of the field of turbulence, it is necessary to give more attention to those statistical quantities which are of special physical importance and which are the working tools of the subject. In addition, we wish to derive all the consequences of the continuity condition and of any symmetry conditions which might be imposed on the turbulence. The analysis in this chapter is exact and deductive; intuitive hypotheses will not make their appearance until we consider the dynamics of turbulent motion. In place of the general discussion, in the preceding chapter, of the joint-probability distribution at arbitrary points, we shall be obliged, on the grounds of expediency, to consider only the case $n=2$. In general, two-point velocity-product mean values of orders two, three and four are the only mean values which find a place in practical theories, although there would be no great difficulty (beyond complexity) in considering the kinematical conditions imposed on higher order mean values.

As before, the axes of reference are such that the fluid has no mean velocity, i.e.

$$\overline{u_i(\mathbf{x})}=0.$$

Constant use will be made of the property of homogeneity in the following typical manner:

$$\overline{u_i(\mathbf{x})\frac{\partial u_i(\mathbf{x})}{\partial x_j}}=\frac{1}{2}\frac{\partial \overline{u_i^2(\mathbf{x})}}{\partial x_j}=0.$$

3.1. The velocity correlation and spectrum tensors

In this section we continue the discussion of § 2.4 in order to exhibit some of the relations concerning the correlation and spectrum tensors which are of particular interest in turbulence theory.

Although three-dimensional Fourier transforms are appropriate to a function of a vector argument, the experimenter can make a Fourier analysis (by passing an electronic signal proportional to the velocity through a filter circuit, or wave analyser) with respect to one space coordinate only.† The resulting spectrum function is a one-dimensional Fourier transform of the velocity correlation tensor, and is obtained from the spectrum tensor $\Phi_{ij}(\boldsymbol{\kappa})$ by integrating over all values of the lateral components of $\boldsymbol{\kappa}$. For instance, if a Fourier analysis of the variation of the velocity along a line in the direction of the coordinate x_1 is made, the resulting spectrum function is

$$\Theta_{ij}(\kappa_1) = \frac{1}{2\pi}\int_{-\infty}^{\infty} R_{ij}(r_1, 0, 0)\, e^{-i\kappa_1 r_1}\, dr_1$$
$$= \int_{-\infty}^{\infty}\int_{-\infty}^{\infty} \Phi_{ij}(\kappa_1, \kappa_2, \kappa_3)\, d\kappa_2\, d\kappa_3. \qquad (3.1.1)$$

When strong symmetry conditions are imposed on the turbulence, the relation between $\Theta_{ij}(\kappa_1)$ and $\Phi_{ij}(\boldsymbol{\kappa})$ becomes very simple, and a knowledge of either one is sufficient to determine the functional form of the other. In the case $i=j=1$, $\Theta_{ij}(\kappa_1)$ becomes a 'longitudinal' one-dimensional spectrum, while $i=j=2$ or 3 gives a 'lateral' spectrum

Correlation and spectrum functions of a single scalar variable can also be obtained by averaging $R_{ij}(\mathbf{r})$ and $\Phi_{ij}(\boldsymbol{\kappa})$ over all directions of the vector arguments $\mathbf{r}$ and $\boldsymbol{\kappa}$. Thus we can define new tensors:

$$S_{ij}(r) = \frac{1}{4\pi r^2}\int R_{ij}(\mathbf{r})\, dA(r), \quad \Psi_{ij}(\kappa) = \int \Phi_{ij}(\boldsymbol{\kappa})\, dA(\kappa), \quad (3.1.2)$$

where $r=|\mathbf{r}|$, $\kappa=|\boldsymbol{\kappa}|$, and the integrations are over the surfaces of spheres of which dA is an element. The insertion of the factor $(4\pi r^2)^{-1}$ and the omission of the corresponding factor $(4\pi\kappa^2)^{-1}$ are

† More precisely, it is possible to record only the variation in the velocity with time at a fixed point as the field of turbulence is carried along by the uniform stream with speed U in the wind tunnel. The frequency analysis of this velocity variation is assumed to be approximately identical with the wave-number analysis of the variation of velocity along a line in the direction of the stream in the corresponding idealized field of homogeneous turbulence. The identification of the two spectra is clearly more accurate for smaller values of $\overline{\mathbf{u}^2}/U^2$, and ample evidence exists for its accuracy at all except the smallest frequencies and wave-numbers under normal experimental conditions. The assumption was introduced first by G. I. Taylor (1938*b*), who demonstrated its accuracy in one particular case by showing that the measured spectrum satisfied the Fourier transform relation with the measured correlation function. C. C. Lin (1950) has estimated the error when the assumption is used in a determination of the dispersion of the spectrum about zero wave-number and finds that it is negligible.

intended to give quantities with clear physical meanings; $S_{ij}(r)$ is the average correlation tensor for two points distance r apart, while $\Psi_{ij}(\kappa)\,d\kappa$ is the contribution to the energy tensor $\overline{u_i u_j}$ from wave-numbers whose magnitudes lie between κ and $\kappa + d\kappa$. The fundamental transform relation (2.4.2) between $R_{ij}(\mathbf{r})$ and $\Phi_{ij}(\boldsymbol{\kappa})$ then shows that

$$S_{ij}(r) = \int \Phi_{ij}(\boldsymbol{\kappa}) \frac{\sin \kappa r}{\kappa r} \,d\boldsymbol{\kappa} = \int_0^\infty \Psi_{ij}(\kappa) \frac{\sin \kappa r}{\kappa r} \,d\kappa, \tag{3.1.3}$$

$$\Psi_{ij}(\kappa) = \frac{\kappa^2}{2\pi^2} \int R_{ij}(\mathbf{r}) \frac{\sin \kappa r}{\kappa r} \,d\mathbf{r} = \frac{2}{\pi} \int_0^\infty S_{ij}(r)\, \kappa r \sin \kappa r \,dr, \tag{3.1.4}$$

i.e. $rS_{ij}(r)$ and $\kappa^{-1}\Psi_{ij}(\kappa)$ are Fourier sine transforms. These tensor functions of a vector magnitude play an important part in the theory of isotropic turbulence, since the dependence on the direction of $\mathbf{r}$ or $\boldsymbol{\kappa}$ is then prescribed by the spherical symmetry.

Of particular physical interest is the *energy spectrum function*

$$E(\kappa) = \tfrac{1}{2}\Psi_{ii}(\kappa), \tag{3.1.5}$$

which is the density of contributions to the kinetic energy on the wave-number magnitude axis (the symbol E is an exception to our convention that Roman letters denote product mean values and Greek letters denote their transforms, but is demanded by common usage). The total kinetic energy per unit mass of fluid is

$$\tfrac{1}{2}\overline{u_i(\mathbf{x})\, u_i(\mathbf{x})} = \int_0^\infty E(\kappa)\,d\kappa. \tag{3.1.6}$$

When hypotheses of statistical equilibrium of the turbulence are introduced in later chapters, it will be found desirable to have available a single quantity which characterizes the part of the turbulence that is associated with a given length scale. The function $E(\kappa)$ fills this need, and is the most important single quantity of the problem—as, indeed, it is in the (kinematically) analogous problems of 'white' light, electrical 'noise' and acoustics.

The behaviour of the spectrum tensor $\Phi_{ij}(\boldsymbol{\kappa})$, and of the associated functions $\Psi_{ij}(\kappa)$ and $E(\kappa)$, at small values of κ can be determined with the aid of Cramér's theorem and the incompressibility condition. We shall need to assume that the first few at least of the derivatives of $\Phi_{ij}(\boldsymbol{\kappa})$ at $\kappa = 0$ exist, which is equivalent to assuming that the corresponding weighted integrals of $R_{ij}(\mathbf{r})$ exist;† stationary random functions do not necessarily satisfy such conditions in general, but it will be shown later that experimental evidence confirms the validity of the assumption in the case of homogeneous turbulence. In the neighbourhood of $\kappa=0$, the spectrum tensor can be written as

$$\Phi_{ij}(\boldsymbol{\kappa})=C_{ij}+\kappa_l C_{ijl}+\kappa_l\kappa_m C_{ijlm}+O(\kappa^3),$$

where the tensor coefficients C_{ij}, C_{ijl}, C_{ijlm} depend on time only. The incompressibility condition (2.4.7) requires

$$\kappa_i C_{ij}+\kappa_i\kappa_l C_{ijl}+\kappa_i\kappa_l\kappa_m C_{ijlm}+O(\kappa^4)=0, \tag{3.1.7}$$

which can be satisfied for all values of $\boldsymbol{\kappa}$ only if

$$C_{ij}=0.$$

Then Cramér's theorem (see § 2.4) establishes that

$$X_i X_j^* \kappa_l C_{ijl} \geqslant 0 \tag{3.1.8}$$

for all sufficiently small κ and arbitrary X_i; since the sign of this expression could be changed by reversing the direction of $\boldsymbol{\kappa}$, the only possibility is

$$C_{ijl}=0.$$

The expression for $\Phi_{ij}(\boldsymbol{\kappa})$ in the neighbourhood of $\kappa=0$ is therefore

$$\Phi_{ij}(\boldsymbol{\kappa})=\kappa_l\kappa_m C_{ijlm}+O(\kappa^3), \tag{3.1.9}$$

where, in view of (3.1.7), the leading coefficient C_{ijlm} satisfies

$$\sum_{\text{perm } i,\, l,\, m} C_{ijlm}=\sum_{\text{perm } j,\, l,\, m} C_{ijlm}=0, \tag{3.1.10}$$

the summations being over all permutations of the indices shown. We note from (3.1.10) that $C_{ijlm}=0$ when $i=j=l=m$, corresponding to the fact that longitudinal 'waves' are not permitted in an incompressible fluid. The behaviour of the wave-number magnitude spectrum defined by (3.1.2) (which is necessarily an even function of κ) is thus

$$\Psi_{ij}(\kappa)=C_{ijlm}\int \kappa_l\kappa_m \mathrm{d}A(\kappa)+O(\kappa^6).$$

$$=\frac{4\pi}{3}C_{ijll}\kappa^4+O(\kappa^6), \tag{3.1.11}$$

† See note on p. 195.

while that of the energy spectrum function $E(\kappa)$ is

$$E(\kappa) = C\kappa^4 + O(\kappa^6), \tag{3.1.12}$$

where the coefficient C is given by

$$C = \frac{2\pi}{3} C_{iill}. \tag{3.1.13}$$

It appears that the density in wave-number space of contributions to the energy tensor $\overline{u_i u_j}$ falls rapidly to zero as the origin is approached. This result is wholly a consequence of the incompressibility condition; if the fluid is compressible, the expansions (3.1.9) and (3.1.12) in general begin with terms of zero and second degree respectively.

3.2. The vorticity correlation and spectrum tensors

It has already been remarked that there is no essential difficulty in the kinematical consideration of mean products of velocity derivatives. There is one particular case which deserves special mention, viz. mean values of products involving the vorticity† vector

$$\boldsymbol{\omega}(\mathbf{x}) = \nabla \times \mathbf{u}(\mathbf{x}),$$

of which the components are‡

$$\omega_i(\mathbf{x}) = \epsilon_{ijk} \frac{\partial u_k(\mathbf{x})}{\partial x_j}.$$

$\boldsymbol{\omega}$ is an axial vector and (unlike the polar vector $\mathbf{u}$) involves a definition of a sense of rotation as well as of the vector defined by the components ω_1, ω_2 and ω_3. The correlation between components of the vorticity at two different points of space, $\mathbf{x}$ and $\mathbf{x}' = \mathbf{x} + \mathbf{r}$, is

$$\begin{aligned} \overline{\omega_i(\mathbf{x})\,\omega_j(\mathbf{x}')} &= \epsilon_{ilm}\epsilon_{jpq} \overline{\frac{\partial u_m(\mathbf{x})}{\partial x_l} \frac{\partial u_q(\mathbf{x}')}{\partial x'_p}} \\ &= -(\delta_{ij}\delta_{lp}\delta_{mq} + \delta_{ip}\delta_{lq}\delta_{mj} + \delta_{iq}\delta_{lj}\delta_{mp} - \delta_{ij}\delta_{lq}\delta_{mp} \\ &\qquad - \delta_{ip}\delta_{lj}\delta_{mq} - \delta_{iq}\delta_{lp}\delta_{mj}) \frac{\partial^2 R_{mq}(\mathbf{r})}{\partial r_l \partial r_p} \\ &= -\delta_{ij}\nabla^2 R_{ll}(\mathbf{r}) + \frac{\partial^2 R_{ll}(\mathbf{r})}{\partial r_i \partial r_j} + \nabla^2 R_{ji}(\mathbf{r}), \end{aligned} \tag{3.2.1}$$

† Turbulent motion is by definition rotational, since the velocity would not otherwise be a random function of position.

‡ ϵ_{ijk} is the unit alternating tensor and has the values $\epsilon_{ijk} = 0$ when i, j and k are not all different, $\epsilon_{ijk} = +1$ or -1 when i, j and k are all different and in cyclic or acyclic order respectively.

in which we have made use of the continuity relation (2.4.6). Contraction of the indices gives the interesting result

$$\overline{\omega_i(\mathbf{x})\,\omega_i(\mathbf{x}+\mathbf{r})} = -\nabla^2 R_{ii}(\mathbf{r}); \tag{3.2.2}$$

which can clearly be generalized to give

$$\overline{[\nabla\times\boldsymbol{\omega}(\mathbf{x})]_i\,[\nabla\times\boldsymbol{\omega}(\mathbf{x}+\mathbf{r})]_i} = -\nabla_r^2\,\overline{\omega_i(\mathbf{x})\,\omega_i(\mathbf{x}+\mathbf{r})} = \nabla^4 R_{ii}(\mathbf{r})$$

and a sequence of such relations.

There is now no difficulty in determining the vorticity spectrum tensor $\Omega_{ij}(\boldsymbol{\kappa})$, defined as the Fourier transform of $\overline{\omega_i(\mathbf{x})\,\omega_j(\mathbf{x}')}$, in terms of the energy spectrum tensor. For, on substituting (2.4.2) in (3.2.1) we find

$$\Omega_{ij}(\boldsymbol{\kappa}) = (\delta_{ij}\kappa^2 - \kappa_i\kappa_j)\,\Phi_{ll}(\boldsymbol{\kappa}) - \kappa^2\Phi_{ji}(\boldsymbol{\kappa}). \tag{3.2.3}$$

Thus the density in wave-number space of contributions to the mean-square total vorticity is

$$\Omega_{ii}(\boldsymbol{\kappa}) = \kappa^2\Phi_{ii}(\boldsymbol{\kappa}), \tag{3.2.4}$$

as is also clear from (3.2.2). It will be found, when we consider the dynamics of the motion, that this spectrum of mean-square total vorticity is identical with the spectrum of the viscous dissipation of kinetic energy. The ratio of the mean-square velocity to the mean-square vorticity evidently provides one of the most important of the lengths characteristic of the turbulence. (3.2.2) and (3.2.4) show that this length is

$$\left[\frac{\overline{u_i(\mathbf{x})\,u_i(\mathbf{x})}}{\overline{\omega_i(\mathbf{x})\,\omega_i(\mathbf{x})}}\right]^{\frac{1}{2}} = \left[-\frac{R_{ii}(\mathbf{r})}{\nabla^2 R_{ii}(\mathbf{r})}\right]^{\frac{1}{2}}_{r=0}$$

$$= \left[\frac{\int \Phi_{ii}(\boldsymbol{\kappa})\,d\boldsymbol{\kappa}}{\int \kappa^2\Phi_{ii}(\boldsymbol{\kappa})\,d\boldsymbol{\kappa}}\right]^{\frac{1}{2}} = \left[\frac{\int_0^\infty E(\kappa)\,d\kappa}{\int_0^\infty \kappa^2 E(\kappa)\,d\kappa}\right]^{\frac{1}{2}}, \tag{3.2.5}$$

that is, it is the reciprocal of the dispersion of the energy about the origin in wave-number space.

3.3. Symmetry conditions

As already suggested, there is some practical value, as well as a gain in simplicity, in a consideration of fields of turbulence which satisfy certain symmetry conditions in a statistical sense. These new conditions must be imposed on the joint-probability distribution of the values of the velocity at any n points of space at a given time of decay. The joint-probability distribution has already been assumed to be independent of arbitrary translations of the configuration formed by the n points and, with the new assumption of symmetry of some kind, it becomes independent also of certain rigid rotations of the configuration relative to the fluid. The axes to which the velocity is referred must rotate with the configuration of n points, and it will therefore be more convenient if we consider the joint-probability distribution of the components, in specified directions, of the velocities at the n points; the symmetry conditions then require the probability distribution to be independent of certain rigid rotations, relative to axes with respect to which the fluid has zero mean motion, of the configuration formed by both the set of n points and the directional vectors specifying the components of the velocity at these points.

In the simple case of turbulence which is spherically symmetrical, the probability distribution is independent of *arbitrary* rigid rotations of the configuration formed by the n points and the various directional vectors. It is possible to go further and require that the probability distribution be also independent of reflexion of the configuration in any plane; this extra condition is included in the usual definition of isotropic turbulence. Various other possibilities exist. The turbulence may have rotational symmetry about a given line, in which case the joint-probability distribution of arbitrary components of the velocities at any n points is independent of arbitrary rigid rotations, about the given line, of the configuration formed by the n points and the associated velocity directions. If, in addition, the probability distribution is independent of reflexion of the configuration in planes through and normal to the axis of symmetry, the turbulence may be said to have axial symmetry. This is a case which also occurs frequently in practice, since many of the experimental devices for generating or modifying a field of turbulence confer a distinction on one particular direction (viz. the direction of the uniform stream carrying the turbulence). The final possibility is that the turbulence has symmetry about a plane only, i.e. that the probability distribution is independent of reflexions of the configuration in the given plane.

The method of establishing the consequences of these symmetry conditions has been expounded by H. P. Robertson (1940). The typical member (of m-order, $m \geqslant n$) of the set of velocity-product mean values derived from the joint-probability distribution of the velocities at n points is displayed in (2.3.1). The corresponding mean value of the product of velocity components in the directions of the m unit vectors $\mathbf{a}, \mathbf{b}, \ldots, \mathbf{h}$, one for each of the m suffixes $i, j, \ldots, p$, is

$$Q(\mathbf{r}, \mathbf{s}, \ldots, \mathbf{a}, \mathbf{b}, \ldots) = \overline{a_i u_i(\mathbf{x}_1)\, b_j u_j(\mathbf{x}_2) \ldots h_p u_p(\mathbf{x}_m)}$$
$$= a_i b_j \ldots h_p Q^{(m)}_{ij\ldots p}(\mathbf{r}, \mathbf{s}, \ldots). \qquad (3.3.1)$$

This scalar quantity Q is fully determined by the $n-1$ vectors $\mathbf{r}, \mathbf{s}, \ldots$ describing the configuration of n points, the m vectors $\mathbf{a}, \mathbf{b}, \ldots$, and the probability laws describing the turbulence. Symmetry conditions provide the information that Q is independent of certain rigid rotations and reflexions of the configuration formed by the vector arguments $\mathbf{r}, \mathbf{s}, \ldots$ and $\mathbf{a}, \mathbf{b}, \ldots$, and the task is to determine the consequent form of $Q^{(m)}_{ij\ldots p}$ consistent with (3.3.1). This is a purely mathematical problem and has applications to fields other than turbulence.

Taking first the case of isotropic turbulence, Q is independent of arbitrary rigid-body rotations and reflexions of the vector configuration; that is, Q is invariant under the full rotation group. Robertson has pointed out that it is a rigorous result of group invariant theory that such an invariant function can be expressed in terms of the fundamental invariants, under the same operation, of the vectors $\mathbf{a}, \mathbf{b}, \ldots$ and $\mathbf{r}, \mathbf{s}, \ldots$ on which Q depends. These fundamental invariants are the scalar products† $a_i b_i$, $a_i r_i$, $r_i s_i$, r^2, s^2, etc., formed from pairs of the various vectors; scalar products like $a_i a_i$ can be ignored since they are unity. In geometrical language, the invariance of the function Q under arbitrary rigid rotations and reflexions of the configuration of vector arguments implies that Q

† Which will not all be independent if more than three vectors are involved.

depends only on the lengths of and angles between these vector arguments of Q. (The determinants or vector triple products like $\epsilon_{ijk}a_i b_j r_k$ are also fundamental invariants under arbitrary rigid rotations of the vector configuration—having the meaning of volumes of the parallelepipeds formed by any three of the vectors—but they change sign when the configuration is reflected in any point and are therefore not isotropic forms. The square of $\epsilon_{ijk}a_i b_j r_k$—and likewise the product of two such expressions—has the required properties, but is itself expressible in terms of the scalar products $a_i b_i$, $b_i r_i$, $r_i a_i$ and r^2. Nevertheless, these skew-isotropic forms are not without their uses, as Robertson (1940) and Chandrasekhar (1950*a*) have shown.) Hence we have

$$Q(\mathbf{r}, \mathbf{s}, \ldots, \mathbf{a}, \mathbf{b}, \ldots) = Q(a_i b_i, a_i r_i, r_i s_i, \ldots, r^2, s^2, \ldots). \quad (3.3.2)$$

But according to (3.3.1) Q is linear in the components of each of the vectors $\mathbf{a}$, $\mathbf{b}$, ... and is a homogeneous function of these vectors. The general expression for Q must therefore be the sum of a number of terms like

$$A(r^2, s^2, r_i s_i, \ldots)(a_i r_i)(b_j s_j)(c_k d_k) \ldots,$$

in which each of the components a_i, b_i, c_i, d_i, ... occurs once only, in a contracted product, with either another member of the same group or one of r_i, s_i, ..., and the scalar coefficient A is an arbitrary function of the scalar products not containing $\mathbf{a}$, $\mathbf{b}$, ... Comparison with (3.3.1) shows that the corresponding general form of $Q^{(m)}_{ij\ldots}$ is

$$Q^{(m)}_{ijkl\ldots}(\mathbf{r}, \mathbf{s}, \ldots) = \Sigma A(r^2, s^2, r_i s_i, \ldots) r_i s_j \delta_{kl} \ldots, \quad (3.3.3)$$

where the summation is over all possible terms such that the suffixes i, j, k, l, ... are attached to any one of the vectors $\mathbf{r}$, $\mathbf{s}$, ... or occur in pairs in the tensor δ_{ij}. The general procedure will be illustrated if we write down the general form of isotropic tensors of the first, second, third and fourth orders which are functions of $\mathbf{r}$ alone (i.e. which involve velocities at two points only):

$$Q_i(\mathbf{r}) = Ar_i, \quad (3.3.4)$$

$$Q_{ij}(\mathbf{r}) = Ar_i r_j + B\delta_{ij}, \quad (3.3.5)$$

$$Q_{ijk}(\mathbf{r}) = Ar_i r_j r_k + Br_i \delta_{jk} + Cr_j \delta_{ki} + Dr_k \delta_{ij}, \quad (3.3.6)$$

$$Q_{ijkl}(\mathbf{r}) = Ar_i r_j r_k r_l + Br_i r_j \delta_{kl} + Cr_j r_k \delta_{li} + Dr_k r_l \delta_{ij} + Er_l r_i \delta_{jk} + Fr_i r_k \delta_{jl} + Gr_j r_l \delta_{ik} + H\delta_{ij}\delta_{kl} + I\delta_{ik}\delta_{jl} + J\delta_{il}\delta_{jk}, \quad (3.3.7)$$

where the scalar coefficients A, B, ... are all even functions of r.

The extension of these results to the case of other, less restrictive, symmetry conditions is very simple. The next simplest case is that of axisymmetric turbulence, when $Q(\mathbf{r}, \mathbf{s}, \ldots, \mathbf{a}, \mathbf{b}, \ldots)$ has to be determined under the condition that it is invariant for rigid-body rotations of the configuration of vector arguments about a given unit vector $\boldsymbol{\lambda}$ and for reflexions of the configuration in any point. In other words, Q is invariant provided $\boldsymbol{\lambda}$ and the vectors $\mathbf{r}, \mathbf{s}, \ldots$, $\mathbf{a}, \mathbf{b}, \ldots$ maintain the same relative configuration, and the condition that $Q(\mathbf{r}, \mathbf{s}, \ldots, \mathbf{a}, \mathbf{b}, \ldots)$ should be an axisymmetric form is identical with the condition that $Q(\boldsymbol{\lambda}, \mathbf{r}, \mathbf{s}, \ldots, \mathbf{a}, \mathbf{b}, \ldots)$ should be an isotropic form. Relaxing the symmetry conditions to permit a dependence of the probability laws on a single direction vector is thus equivalent, so far as the general form of $Q^{(m)}_{ij\ldots}$ is concerned, to increasing by one the number of space vectors on which $Q^{(m)}_{ij\ldots}$ depends (with the slight difference that the dependence of the scalar coefficients $A, B, \ldots$ on $\lambda_i \lambda_i = 1$ is of no significance). Thus the general forms of axisymmetric tensors of the first, second and third order which involve velocities at two points only are:

$$Q_i(\mathbf{r}) = Ar_i + B\lambda_i, \tag{3.3.8}$$

$$Q_{ij}(\mathbf{r}) = Ar_i r_j + B\lambda_i \lambda_j + C\delta_{ij} + Dr_i \lambda_j + Er_j \lambda_i, \tag{3.3.9}$$

$$\begin{aligned} Q_{ijk}(\mathbf{r}) = {} & Ar_i r_j r_k + B\lambda_i \lambda_j \lambda_k + Cr_i \delta_{jk} + Dr_j \delta_{ki} + Er_k \delta_{ij} \\ & + F\lambda_i \delta_{jk} + G\lambda_j \delta_{ki} + H\lambda_k \delta_{ij} + Ir_i r_j \lambda_k + Jr_j r_k \lambda_i \\ & + Kr_k r_i \lambda_j + Lr_i \lambda_j \lambda_k + Mr_j \lambda_k \lambda_i + Nr_k \lambda_i \lambda_j, \end{aligned} \tag{3.3.10}$$

where the scalar functions $A, B, \ldots$ are arbitrary functions of r^2 and $r_i \lambda_i$.

In the case of turbulence which has spherical symmetry but does not have reflexional symmetry, the general form of Q includes terms made up from additional fundamental invariants like $\epsilon_{ijk} a_i b_j r_k$ which change sign on reflexion. The general form of $Q^{(m)}_{ij\ldots}(\mathbf{r}, \mathbf{s}, \ldots)$ is likewise modified, and the summation in (3.3.3) must be extended to include a number of new terms. It is not necessary to go into the details, which are quite straightforward; for example, (3.3.4) is unchanged and (3.3.5) is replaced by

$$Q_{ij}(\mathbf{r}) = Ar_i r_j + B\delta_{ij} + C\epsilon_{ijk} r_k. \tag{3.3.11}$$

The case of turbulence with rotational symmetry about a given direction $\boldsymbol{\lambda}$ (but without reflexional symmetry) is now obtained from this case of spherical symmetry by permitting the dependence on an extra vector $\boldsymbol{\lambda}$, just as axisymmetry can be considered as equivalent to isotropy with an extra vector argument.

The least restrictive symmetry condition is that of symmetry about a plane. The general forms of the tensors in this case can be obtained from those for isotropic turbulence with a dependence on two extra directional vectors; for instance, the expression for $Q_{ij}(\mathbf{r})$ is the next member of the sequence established by (3.3.5) and (3.3.9). Even when no symmetry conditions at all are imposed the form of $Q^{(m)}_{ij\ldots}(\mathbf{r}, \mathbf{s}, \ldots)$ is not completely arbitrary, since it must transform as a tensor. This case of no symmetry conditions is equivalent to the case of spherical symmetry (without reflexional symmetry) with two extra (orthogonal) vector arguments, $\boldsymbol{\lambda}$ and $\boldsymbol{\mu}$ say, so that, on extending (3.3.11) appropriately, we find that the most general expression for $Q_{ij}(\mathbf{r})$ is

$$Q_{ij}(\mathbf{r}) = A\delta_{ij} + (3 \text{ terms like } Br_i r_j) + (6 \text{ like } Cr_i\lambda_j) + (3 \text{ like } D\epsilon_{ijk}r_k) + (12 \text{ like } E\epsilon_{ikl}r_k\lambda_l r_j) + (6 \text{ like } F\epsilon_{ikl}r_k\lambda_l\mu_j), \quad (3.3.12)$$

where the scalar coefficients $A, B, \ldots$ are functions of r^2, $r_i\lambda_i$ and $r_i\mu_i$. The relative simplicity of the expression (3.3.5) for the corresponding isotropic tensor is considerable!

It frequently happens that the above general expressions are simplified by the existence of symmetry in the suffixes; for instance, $\overline{u_i(\mathbf{x})\,u_j(\mathbf{x})\,u_k(\mathbf{x}+\mathbf{r})}$ is unchanged by interchange of the suffixes i and j. The need to satisfy the continuity condition will simplify many of the above general expressions still further. The incompressibility relation (2.3.4) shows that $Q^{(m)}_{ij\ldots}(\mathbf{r}, \mathbf{s}, \ldots)$ is solenoidal in all those indices that are alone in referring to the velocity at one of the n points of the configuration—for instance, $\overline{u_i(\mathbf{x})\,u_j(\mathbf{x})\,u_k(\mathbf{x}+\mathbf{r})}$ is solenoidal in k (with respect to $\mathbf{r}$) but not in i or j—and this condition of vanishing divergence will require the scalar functions occurring in the expression for $Q^{(m)}_{ij\ldots}$ to satisfy certain differential relations. Chandrasekhar (1950*a*), following Robertson (1940), has pointed out that the process of introducing the scalar functions and then eliminating some of them with the aid of the continuity relations can be avoided by writing $Q^{(m)}_{ij\ldots}$, from the beginning, as the curl (with respect to the index in which $Q^{(m)}_{ij\ldots}$ is solenoidal) of another tensor of order m. In this way the continuity condition is satisfied identically. Since the operation of taking the curl does not preserve reflexional symmetry, the problem of determining the general form of, say, an isotropic tensor of order m which is solenoidal in one index is reduced to that of finding a suitable form (containing the requisite number of scalar functions) of a skew-isotropic tensor of order m. The technique is extremely useful in problems involving complicated tensor quantities, but we shall not need to use it in this work.

Finally, we may note that the general forms, demanded by the various symmetry conditions, of mean values involving velocity derivatives and of Fourier transforms of velocity-product mean values can be obtained by making appropriate operations on the above results, or they can be obtained directly by means of the same procedure. The presence of a velocity derivative adds a unit vector (specifying the direction of the derivative) to the configuration on which the mean value depends, while in the case of a Fourier transform the vector arguments $\mathbf{r}, \mathbf{s}, \ldots$ are replaced by the corresponding wave-number vectors.

3.4. Isotropic turbulence

We consider now the very important special case of isotropic turbulence. The velocity correlation $R_{ij}(\mathbf{r})$ is then an isotropic second-order two-point tensor and therefore has the form established in the preceding section (see (3.3.5)), viz.

$$R_{ij}(\mathbf{r}) = F(r)\, r_i r_j + G(r)\, \delta_{ij}, \tag{3.4.1}$$

where F and G are arbitrary scalar functions of r^2 (and also of t, but we are ignoring the dependence on time until the dynamics of the motion are considered). The condition of isotropy has already ensured that $R_{ij}(\mathbf{r})$ is symmetrical in the two suffixes. The continuity condition (2.4.6) must also be satisfied, and we find that

$$\frac{\partial R_{ij}(\mathbf{r})}{\partial r_i} = r_j\left(4F + r\frac{\partial F}{\partial r} + \frac{1}{r}\frac{\partial G}{\partial r}\right) = 0.$$

This equation is to be satisfied for all values of r and consequently

$$4F + r\frac{\partial F}{\partial r} + \frac{1}{r}\frac{\partial G}{\partial r} = 0, \tag{3.4.2}$$

showing that only one scalar function is needed to specify the velocity correlation when the turbulence is isotropic.

It is found more convenient, in connexion with experimental work, to introduce an alternative pair of scalar functions defined by

$$f(r)=\frac{\overline{u_p(\mathbf{x})\,u_p(\mathbf{x}+\mathbf{r})}}{\overline{u_p^2}},\quad g(r)=\frac{\overline{u_n(\mathbf{x})\,u_n(\mathbf{x}+\mathbf{r})}}{\overline{u_n^2}},\tag{3.4.3}$$

where u_p and u_n denote velocity components parallel and normal respectively to the vector separation $\mathbf{r}$ (it should be noted that p and n are not tensor suffixes, so that the summation convention does not operate). That is, $f(r)$ is the *longitudinal* velocity correlation coefficient and $g(r)$ is the *lateral* velocity correlation coefficient for two points at distance r apart in any direction (see fig. 3.1). These

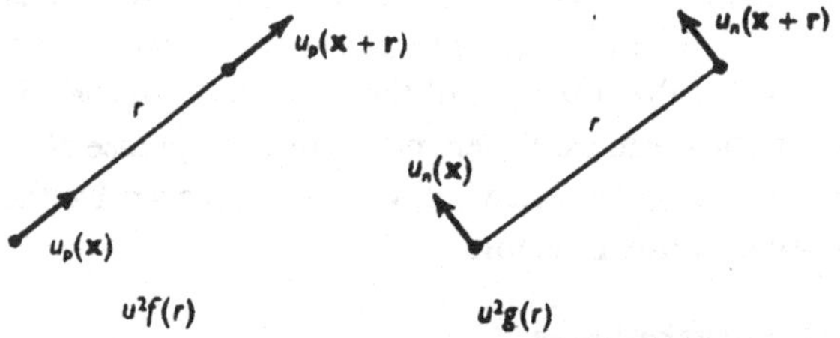

Fig. 3.1. Longitudinal and lateral velocity correlations.

two quantities can readily be measured in a wind-tunnel stream. Their relation to the functions F and G is found from the general form (3.4.1) to be

$$\left.\begin{aligned}\overline{u_p(\mathbf{x})\,u_p(\mathbf{x}+\mathbf{r})}&=r^2F(r)+G(r)=u^2f(r),\\ \overline{u_n(\mathbf{x})\,u_n(\mathbf{x}+\mathbf{r})}&=G(r)=u^2g(r),\end{aligned}\right\}\tag{3.4.4}$$

where u is used here and elsewhere to denote the root-mean-square of any velocity component, i.e.

$$u^2=\overline{u_p^2}=\overline{u_n^2}=\tfrac{1}{3}\overline{u_iu_i}.$$

In terms of f and g, the relations (3.4.1) and (3.4.2) become

$$R_{ij}(\mathbf{r})=u^2\left(\frac{f-g}{r^2}r_ir_j+g\delta_{ij}\right),\tag{3.4.5}$$

$$g=f+\tfrac{1}{2}rf',\tag{3.4.6}$$

where $f'=\partial f/\partial r$.

One or two general features of the functions $f(r)$ and $g(r)$ are immediately evident. From (3.4.3) we find

$$f(0) = g(0) = 1,$$

and the general Schwarz inequality mentioned in § 2.4 shows that this is the maximum value of f and g. Assuming that the second derivative of $f(r)$ at $r=0$ exists (which will be seen to require only that the rate of energy dissipation by viscous forces be finite), the behaviour in the neighbourhood of the maximum is

$$f(r) = 1 + \tfrac{1}{2} f_0'' r^2 + O(r^4).$$

It is common practice to introduce the length λ defined by

$$f_0'' = -\frac{1}{\lambda^2},$$

so that near the origin we have (from (3.4.6))

$$f(r) \approx 1 - \frac{r^2}{2\lambda^2}, \quad g(r) \approx 1 - \frac{r^2}{\lambda^2}. \tag{3.4.7}$$

From (3.2.2) (with $r=0$) the mean-square vorticity is

$$\overline{\omega_i \omega_i} = -(\nabla^2 R_{ii})_{r=0} = -15 u^2 f_0'' = \frac{15 u^2}{\lambda^2}, \tag{3.4.8}$$

and it follows from the sequence of relations of which (3.2.2) is the first that the terms in the expansion of $u^2 f(r)$ (or $u^2 g(r)$ or $R_{ii}(\mathbf{r})$) in powers of r^2 are proportional to mean energy, mean-square vorticity, mean square of $\nabla \times (\nabla \times \mathbf{u})$, etc.

As mentioned earlier we shall assume (until experimental results reveal an inconsistency) that f decreases to zero, as $r \to \infty$, with sufficient rapidity to make $\int_0^\infty r^m f(r)\, dr$ convergent for the various values of m that arise in the analysis. The relation between the moments of $f(r)$ and $g(r)$ is found from (3.4.6) to be

$$\int_0^\infty r^m g(r)\, dr = \left(\frac{1-m}{2}\right) \int_0^\infty r^m f(r)\, dr \quad (m \geqslant 0). \tag{3.4.9}$$

The lengths $L_p = \int_0^\infty f(r)\, dr$ and $L_n = \int_0^\infty g(r)\, dr = \tfrac{1}{2} L_p$ are convenient measures of the linear extent of the region within which velocities are appreciably correlated, and are known as the longitudinal and lateral integral scales. For $m = 1$, (3.4.9) becomes

$$\int_0^\infty r g(r)\, dr = 0;$$

and for $m > 1$ the moments of f and g have opposite signs, which suggests that, for large values of r, f is positive and g is negative. (A demonstration that $f(r) > 0$ for all values of r has never been given, but it is a very plausible result for an incompressible fluid and is consistent with all measurements of $f(r)$.) The rough picture of the curves $f(r)$ and $g(r)$ which emerges from all these remarks is shown in fig. 3.2; the experimental curves are found to have this same general shape.

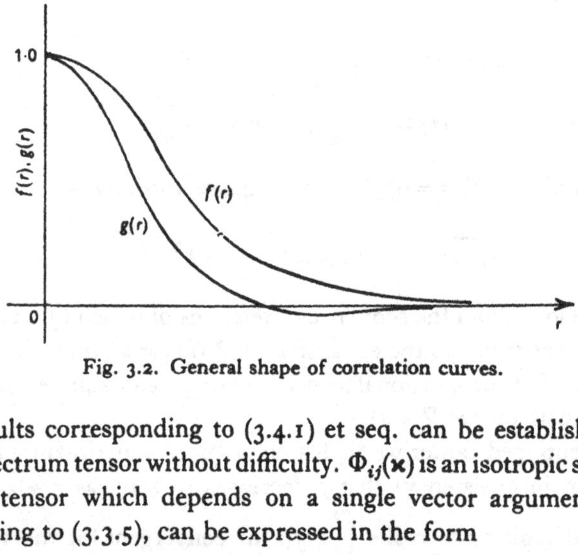

Fig. 3.2. General shape of correlation curves.

Results corresponding to (3.4.1) et seq. can be established for the spectrum tensor without difficulty. $\Phi_{ij}(\boldsymbol{\kappa})$ is an isotropic second-order tensor which depends on a single vector argument and, according to (3.3.5), can be expressed in the form

$$\Phi_{ij}(\boldsymbol{\kappa}) = A(\kappa)\,\kappa_i\kappa_j + B(\kappa)\,\delta_{ij}, \qquad (3.4.10)$$

where A and B are arbitrary even functions of κ. The continuity condition (2.4.7) then requires

$$B = -\kappa^2 A.$$

The density in wave-number space of contributions to the total energy is

$$\tfrac{1}{2}\Phi_{ii}(\boldsymbol{\kappa}) = \tfrac{1}{2}(\kappa^2 A + 3B) = -\kappa^2 A(\kappa),$$

and is a function of the wave-number magnitude alone. The contribution to total energy from that part of wave-number space between spheres of radii κ and $\kappa + d\kappa$ is $E(\kappa)\, d\kappa$ (cf. (3.1.2) and (3.1.5)), where

$$E(\kappa) = 4\pi\kappa^2 . \tfrac{1}{2}\Phi_{ii}(\boldsymbol{\kappa}) = -4\pi\kappa^4 A(\kappa), \tag{3.4.11}$$

and we shall use $E(\kappa)$ as the single scalar function defining $\Phi_{ij}(\boldsymbol{\kappa})$. The expression (3.4.10) for the spectrum tensor becomes

$$\Phi_{ij}(\boldsymbol{\kappa}) = \frac{E(\kappa)}{4\pi\kappa^4}(\kappa^2\delta_{ij} - \kappa_i\kappa_j). \tag{3.4.12}$$

This may be compared with the expression (2.4.10) established for homogeneous turbulence in general. The vorticity spectrum tensor is found from (3.2.3) to be

$$\Omega_{ij}(\boldsymbol{\kappa}) = \frac{E(\kappa)}{4\pi\kappa^2}(\kappa^2\delta_{ij} - \kappa_i\kappa_j) = \kappa^2\Phi_{ij}(\boldsymbol{\kappa}).$$

The relation between the single scalar function $E(\kappa)$ which determines the spectrum tensor and the function which determines the correlation tensor is most easily obtained from the Fourier transform relation (2.4.3) by putting $i = j$ and summing:

$$\Phi_{ii}(\boldsymbol{\kappa}) = \frac{E(\kappa)}{2\pi\kappa^2} = \frac{1}{8\pi^3}\int R_{ii}(\mathbf{r})\, e^{-i\boldsymbol{\kappa}.\mathbf{r}}\, d\mathbf{r}. \tag{3.4.13}$$

But from (3.4.5)

$$R_{ii}(\mathbf{r}) = u^2(f + 2g) = u^2(3f + rf'), \quad = 2R(r), \text{ say}, \tag{3.4.14}$$

whence (3.4.13) and the corresponding inverse relation obtained from (2.4.2) become

$$\left.\begin{aligned} E(\kappa) &= \frac{2}{\pi}\int_0^\infty R(r)\,\kappa r \sin\kappa r\, dr, \\ R(r) &= \int_0^\infty E(\kappa)\frac{\sin\kappa r}{\kappa r}\, d\kappa. \end{aligned}\right\} \tag{3.4.15}$$

The relations (3.4.15) are identical with the relations (3.1.3) and (3.1.4) between the spectrum and correlation tensors averaged over all directions of the vector arguments; when the turbulence is isotropic and when the two suffixes are equalized and summed, no directional averaging process is necessary. The direct relations between $u^2f(r)$ and $E(\kappa)$ are found from (3.4.14) and (3.4.15) to be

$$\left.\begin{aligned} E(\kappa) &= \frac{1}{\pi}\int_0^\infty u^2f(r)\,\kappa^2r^2\left(\frac{\sin\kappa r}{\kappa r}-\cos\kappa r\right)dr,\\ u^2f(r) &= 2\int_0^\infty E(\kappa)\,\kappa^{-2}r^{-2}\left(\frac{\sin\kappa r}{\kappa r}-\cos\kappa r\right)d\kappa. \end{aligned}\right\} \qquad (3.4.16)$$

The longitudinal one-dimensional spectrum function (see (3.1.1)) becomes

$$\begin{aligned} \phi(\kappa_1) = \Theta_{11}(\kappa_1, 0, 0) &= \frac{1}{2\pi}\int_{-\infty}^\infty u^2f(r_1)\cos\kappa_1r_1\,dr_1\\ &= \int_{-\infty}^\infty\int_{-\infty}^\infty \Phi_{11}(\kappa_1, \kappa_2, \kappa_3)\,d\kappa_2\,d\kappa_3\\ &= \frac{1}{2}\int_{\kappa_1}^\infty\left(1-\frac{\kappa_1^2}{\kappa^2}\right)\frac{E(\kappa)}{\kappa}\,d\kappa, \end{aligned} \qquad (3.4.17)$$

and the corresponding explicit expression for $E(\kappa)$ in terms of the measurable function $\phi(\kappa_1)$ is

$$E(\kappa) = \kappa^3\frac{d}{d\kappa}\left[\frac{1}{\kappa}\frac{d\phi(\kappa)}{d\kappa}\right]. \qquad (3.4.18)$$

Similarly, the lateral one-dimensional spectrum function is

$$\begin{aligned} \Theta_{22}(\kappa_1, 0, 0) &= \frac{1}{2\pi}\int_{-\infty}^\infty u^2g(r_1)\cos\kappa_1r_1\,dr_1\\ &= \int_{-\infty}^\infty\int_{-\infty}^\infty \Phi_{22}(\kappa_1, \kappa_2, \kappa_3)\,d\kappa_2\,d\kappa_3\\ &= \frac{1}{4}\int_{\kappa_1}^\infty\left(1+\frac{\kappa_1^2}{\kappa^2}\right)\frac{E(\kappa)}{\kappa}\,d\kappa\\ &= \tfrac{1}{2}\phi(\kappa_1) - \tfrac{1}{2}\kappa_1\frac{d\phi(\kappa_1)}{d\kappa_1}. \end{aligned} \qquad (3.4.19)$$

The relation between the integral moments of $R(r)$ (or $f(r)$) and derivatives of $E(\kappa)$ at $\kappa = 0$ follows from (3.4.15):

$$\begin{aligned} \left(\frac{\partial^{2m}E(\kappa)}{\partial\kappa^{2m}}\right)_{\kappa=0} &= \frac{4m}{\pi}(-1)^{m+1}\int_0^\infty r^{2m}R(r)\,dr\\ &= \frac{4m}{\pi}(m-1)(-1)^m u^2\int_0^\infty r^{2m}f(r)\,dr. \end{aligned} \qquad (3.4.20)$$

The two integral moments of $f(r)$ not given by formula (3.4.20) are readily found from (3.4.15):

$$\int_0^\infty R(r)\,dr = u^2 \int_0^\infty f(r)\,dr = \frac{\pi}{2}\int_0^\infty \kappa^{-1}E(\kappa)\,d\kappa \tag{3.4.21}$$

and

$$u^2 \int_0^\infty r^2 f(r)\,dr = \pi \int_0^\infty \kappa^{-3}E(\kappa)\,d\kappa. \tag{3.4.22}$$

From (3.4.20) we find that the value of the coefficient of κ^4 in the expansion of $E(\kappa)$ in powers of κ (see (3.1.12)) is

$$C = \frac{1}{3\pi}u^2 \int_0^\infty r^4 f(r)\,dr. \tag{3.4.23}$$

In fact, we can go further and obtain, from a comparison of (3.4.12) and the relation (3.1.9) for general homogeneous turbulence,

$$C_{ijlm} = \frac{C}{4\pi}(\delta_{ij}\delta_{lm} - \tfrac{1}{2}\delta_{il}\delta_{jm} - \tfrac{1}{2}\delta_{im}\delta_{jl}). \tag{3.4.24}$$

The integral moments of $E(\kappa)$ and the derivatives of $R(r)$ (or $f(r)$) at $r=0$ are likewise related by

$$\left[\frac{\partial^{2m}R(r)}{\partial r^{2m}}\right]_{r=0} = \tfrac{1}{2}(2m+3)\,u^2\left[\frac{\partial^{2m}f(r)}{\partial r^{2m}}\right]_{r=0} = \frac{(-1)^m}{2m+1}\int_0^\infty \kappa^{2m}E(\kappa)\,d\kappa. \tag{3.4.25}$$

In particular,

$$-f_0'' = \frac{1}{\lambda^2} = \frac{2}{15u^2}\int_0^\infty \kappa^2 E(\kappa)\,d\kappa = \frac{1}{5}\frac{\int_0^\infty \kappa^2 E(\kappa)\,d\kappa}{\int_0^\infty E(\kappa)\,d\kappa}, \tag{3.4.26}$$

showing how the dispersion of the energy in wave-number space determines the radius of curvature of the correlation curves at $r=0$.

The discussion of first- and third-order isotropic tensors proceeds along similar lines, although we shall not need to go into as much detail as in the important case of the second-order tensor. The general form of a first-order isotropic tensor depending on a single vector $\mathbf{r}$ (an example is $\overline{p(\mathbf{x})\,u_i(\mathbf{x}+\mathbf{r})}$) was found to be (see (3.3.4))

$$Q_i(\mathbf{r}) = Ar_i,$$

where A is an arbitrary even function of r. If this tensor is solenoidal (as is $\overline{p(\mathbf{x})\,u_i(\mathbf{x}+\mathbf{r})}$)

$$\frac{\partial Q_i(\mathbf{r})}{\partial r_i}=3A+r\frac{\partial A}{\partial r}=0,$$

which has no non-zero solution which makes $A(r)$ regular at $r=0$. Excluding the possibility that non-regular solutions occur in turbulence theory, we have the important conclusion that solenoidal isotropic vector functions of $\mathbf{r}$ are identically zero.

The most important example of a third-order tensor depending on a single vector $\mathbf{r}$ is $\overline{u_i(\mathbf{x})\,u_j(\mathbf{x})\,u_l(\mathbf{x}+\mathbf{r})}=S_{ijl}(\mathbf{r})$, for which the general form in isotropic turbulence is (see (3.3.6))

$$S_{ijl}(\mathbf{r})=Ar_ir_jr_l+B(r_i\delta_{jl}+r_j\delta_{il})+Dr_l\delta_{ij}, \qquad (3.4.27)$$

in view of the symmetry in the suffixes i and j. The continuity condition requires

$$\frac{\partial S_{ijl}(\mathbf{r})}{\partial r_l}=\left(5A+r\frac{\partial A}{\partial r}+\frac{2}{r}\frac{\partial B}{\partial r}\right)r_ir_j+\left(2B+3D+r\frac{\partial D}{\partial r}\right)\delta_{ij}=0$$

for all values of $\mathbf{r}$, so that the scalar functions A, B, D are related by the two equations

$$5A+rA'+\frac{2}{r}B'=0, \quad 2B+3D+rD'=0. \qquad (3.4.28)$$

An integral relation (which is not independent of (3.4.28)) follows from the fact that $S_{iil}(\mathbf{r})$ is a solenoidal first-order isotropic tensor and is therefore identically zero, that is,

$$r^2A+2B+3D=0. \qquad (3.4.29)$$

From this and the second of the equations (3.4.28) we find

$$A=\frac{1}{r}D', \quad B=-\tfrac{3}{2}D-\tfrac{1}{2}rD', \qquad (3.4.30)$$

showing that $S_{ijl}(\mathbf{r})$, like $R_{ij}(\mathbf{r})$, is determined by a single scalar function of r when the turbulence is isotropic.

The third-order velocity-product mean values, or 'triple-velocity correlations' that are usually measured are those for which the velocity components are either parallel or normal to $\mathbf{r}$. The three possibilities that are independent, so far as the requirement of isotropy alone is concerned, are shown in fig. 3.3 and are commonly expressed in the following notation:

$$\left.\begin{aligned} u^3k(r) &= \overline{u_p^2(\mathbf{x})\,u_p(\mathbf{x}+\mathbf{r})} = \frac{r_i r_j r_l}{r^3} S_{ijl}(\mathbf{r}) = r(r^2A+2B+D) = -2rD, \\ u^3h(r) &= \overline{u_n^2(\mathbf{x})\,u_p(\mathbf{x}+\mathbf{r})} = \frac{n_i n_j r_l}{r} S_{ijl}(\mathbf{r}) = rD, \\ u^3q(r) &= \overline{u_p(\mathbf{x})\,u_n(\mathbf{x})\,u_n(\mathbf{x}+\mathbf{r})} = \frac{r_i n_j n_l}{r} S_{ijl}(\mathbf{r}) = rB = -\tfrac{3}{2}rD - \tfrac{1}{2}r^2D', \end{aligned}\right\}$$

(3.4.31)

where $\mathbf{n}$ is a unit vector normal to $\mathbf{r}$. Of these three mean values, the first is the most easily measurable (since it is easier to measure

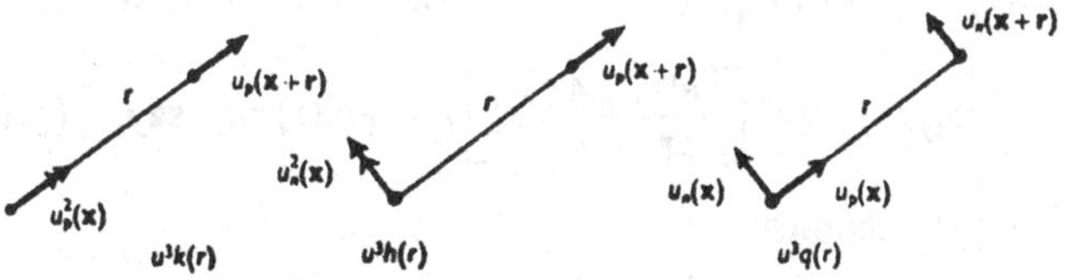

Fig. 3.3. Triple-velocity correlations.

the turbulent-velocity component parallel to the direction of the wind-tunnel stream than to measure the lateral component, and $u^3k(r)$ is the only mean value involving velocity components in a single direction), and we shall adopt $k(r)$ as the single scalar function determining the triple-velocity correlations. The expression for $S_{ijl}(\mathbf{r})$ then becomes

$$S_{ijl}(\mathbf{r}) = u^3\left[\left(\frac{k-rk'}{2r^3}\right) r_i r_j r_l + \left(\frac{2k+rk'}{4r}\right)(r_i\delta_{jl}+r_j\delta_{il}) - \frac{k}{2r} r_l \delta_{ij}\right]. \tag{3.4.32}$$

The functions A, B, D are even in r, while k, h, q are odd. Furthermore, k, h, q are of order r^3 when r is small, since, if we choose $\mathbf{r}$ in the direction of the x_1-axis, we find

$$u^3\left(\frac{\partial k}{\partial r}\right)_{r=0} = \overline{u_1^2(\mathbf{x})\frac{\partial u_1(\mathbf{x})}{\partial x_1}} = \frac{1}{3}\frac{\partial \overline{u_1^3}}{\partial x_1} = 0.$$

The Fourier transform of $S_{ijl}(\mathbf{r})$ defined by

$$\Upsilon_{ijl}(\boldsymbol{\kappa}) = \frac{1}{8\pi^3}\int S_{ijl}(\mathbf{r})\, e^{-\iota\boldsymbol{\kappa}.\mathbf{r}}\, d\mathbf{r}, \tag{3.4.33}$$

will also have the general form (3.3.6), is symmetrical in the indices i and j, and must satisfy the continuity condition

$$\kappa_l \Upsilon_{ijl}(\boldsymbol{\kappa}) = 0.$$

Consequently the expression for $\Upsilon_{ijl}(\boldsymbol{\kappa})$ in isotropic turbulence is

$$\Upsilon_{ijl}(\boldsymbol{\kappa}) = \iota\Upsilon(\kappa)\,(\kappa_i\kappa_j\kappa_l - \tfrac{1}{2}\kappa^2\kappa_i\delta_{jl} - \tfrac{1}{2}\kappa^2\kappa_j\delta_{il}), \tag{3.4.34}$$

where $\Upsilon(\kappa)$ is an arbitrary even scalar function of κ. The relation between the single scalar function $\Upsilon(\kappa)$ defining $\Upsilon_{ijl}(\boldsymbol{\kappa})$ and the single scalar function $k(r)$ defining $S_{ijl}(\mathbf{r})$ is most easily established by putting $i = l$ in (3.4.32) and summing over all values of i; we have

$$S_{iji}(\mathbf{r}) = \tfrac{1}{2}u^3\left[\frac{\partial k(r)}{\partial r} + \frac{4}{r}k(r)\right] r_j = \tfrac{1}{2}K(r)\, r_j, \quad \text{say}, \tag{3.4.35}$$

and (3.4.33) becomes

$$\begin{aligned}
\Upsilon_{iji}(\boldsymbol{\kappa}) &= \frac{1}{16\pi^3}\int K(r)\, r_j\, e^{-\iota\boldsymbol{\kappa}.\mathbf{r}}\, d\mathbf{r} \\
&= \frac{\iota}{16\pi^3}\frac{\partial}{\partial\kappa_j}\left[\int K(r)\, e^{-\iota\boldsymbol{\kappa}.\mathbf{r}}\, d\mathbf{r}\right] \\
&= \frac{\iota}{4\pi^2}\frac{\partial}{\partial\kappa_j}\left[\int_0^\infty K(r)\, r^2 \frac{\sin\kappa r}{\kappa r}\, dr\right] \\
&= -\frac{\iota}{4\pi^2}\frac{\kappa_j}{\kappa^2}\int_0^\infty \frac{\partial(r^3K)}{\partial r}\frac{\sin\kappa r}{\kappa r}\, dr,
\end{aligned}$$

whence comparison with (3.4.34) gives

$$\Upsilon(\kappa) = \frac{1}{4\pi^2}\frac{1}{\kappa^6}\int_0^\infty \left(r\frac{\partial}{\partial r} + 3\right) K(r)\,\kappa r \sin\kappa r\, dr. \tag{3.4.36}$$

The transform of this relation is

$$\left(r\frac{\partial}{\partial r} + 3\right) K(r) = 8\pi\int_0^\infty \kappa^6\Upsilon(\kappa)\frac{\sin\kappa r}{\kappa r}\, d\kappa. \tag{3.4.37}$$

CHAPTER IV

SOME LINEAR PROBLEMS

Before examining the very difficult non-linear problem presented by the decay of homogeneous turbulence, we shall look at some particular problems which are tractable. The common element of these problems is the linearity of the relations involved. This linearity is achieved in almost every case by assuming that the turbulent velocity is suitably small, but the approximation is not so drastic as to rob the problems of any practical interest. In particular, the solution of the linearized problem of the effect of passing homogeneous turbulence through wire gauze—discussed in § 4.2—has received strong support from experimental data.

4.1. Simple harmonic oscillator subject to a random force

There are several different physical problems that can be included within the above heading, and we shall write down the equations without initial reference to any particular situation. We consider a function $X(t)$ of a single variable t which satisfies the differential equation

$$\ddot{X} + \lambda \dot{X} + n^2 X = u(t), \tag{4.1.1}$$

where $\dot{X} = \mathrm{d}X/\mathrm{d}t$, λ and n^2 are positive real constants, and $u(t)$ is a stationary random function of t. This is the usual equation for the displacement of a body, the free motion of which consists of damped simple harmonic one-dimensional oscillations with angular frequency $(n^2 - \frac{1}{4}\lambda^2)^{\frac{1}{2}}$, subject to a random external force proportional to $u(t)$. The variation of $X(t)$ will likewise be random, and the object is to determine the statistical behaviour of $X(t)$ in terms of that of $u(t)$. When $u(t)$ is periodic in t we have the well-known resonance solution for $X(t)$. Since the relation between $X(t)$ and $u(t)$ is linear, it is clear that a stationary random variation of $u(t)$ will produce a similar variation of $X(t)$ and that the method of solution of (4.1.1) lies in resolution of $u(t)$ and $X(t)$ into Fourier components in the manner of (2.5.3).

We write

$$u(t) = \int e^{\iota\kappa t} \, \mathrm{d}Z(\kappa), \quad X(t) = \int e^{\iota\kappa t} \, \mathrm{d}W(\kappa), \tag{4.1.2}$$

where κ is a frequency variable, and increments in the random functions $Z(\kappa)$ and $W(\kappa)$ have the properties

$$\left.\begin{aligned}\lim_{d\kappa_1=d\kappa_2=d\kappa\to 0}\frac{\overline{dZ^*(\kappa_1)\,dZ(\kappa_2)}}{d\kappa}&=0,\\ \lim_{d\kappa_1=d\kappa_2=d\kappa\to 0}\frac{\overline{dW^*(\kappa_1)\,dW(\kappa_2)}}{d\kappa}&=0\end{aligned}\right\}\quad(\kappa_1\neq\kappa_2),$$

$$\lim_{d\kappa\to 0}\frac{\overline{dZ^*(\kappa)\,dZ(\kappa)}}{d\kappa}=\Phi(\kappa),\quad \lim_{d\kappa\to 0}\frac{\overline{dW^*(\kappa)\,dW(\kappa)}}{d\kappa}=\Psi(\kappa),\tag{4.1.3}$$

where $\Phi(\kappa)$ and $\Psi(\kappa)$ are the spectrum (density) functions of $u(t)$ and $X(t)$ respectively. On substituting (4.1.2) in (4.1.1), we find

$$(-\kappa^2+n^2+\iota\kappa\lambda)\,dW(\kappa)=dZ(\kappa).$$

The relation between the spectra of the displacement of the body and of the impressed force is therefore

$$\Psi(\kappa)=\frac{1}{(n^2-\kappa^2)^2+\kappa^2\lambda^2}\Phi(\kappa),\tag{4.1.4}$$

as was to be expected from resonance theory.

A quantity which is often more readily observable is the correlation between the displacements of the body at two different times. It has been shown in Chapter II that this correlation is simply the Fourier transform of the spectrum function; that is,

$$\begin{aligned}\overline{X(t)X(t+\tau)}&=\int_{-\infty}^{\infty}\Psi(\kappa)\,e^{\iota\kappa\tau}\,d\kappa\\ &=\int_{-\infty}^{\infty}\frac{1}{(n^2-\kappa^2)^2+\kappa^2\lambda^2}\Phi(\kappa)\,e^{\iota\kappa\tau}\,d\kappa.\end{aligned}\tag{4.1.5}$$

We can make use of Parseval's formula† to relate this integral to the

† See E. C. Titchmarsh, *Introduction to the theory of Fourier integrals*, Oxford University Press, 1937. One form of Parseval's formula states that if $F(x)$ and $G(x)$ are functions belonging to $L(-\infty,\infty)$ such that

$$\int_{-\infty}^{\infty}f(\kappa)\,e^{\iota\kappa x}\,d\kappa=F(x)\quad\text{and}\quad\int_{-\infty}^{\infty}g(\kappa)\,e^{\iota\kappa x}\,d\kappa=G(x),$$

the Fourier transform of the product $f(\kappa).g(\kappa)$ is

$$\int_{-\infty}^{\infty}f(\kappa)\,g(\kappa)\,e^{\iota\kappa x}\,d\kappa=\frac{1}{2\pi}\int_{-\infty}^{\infty}F(y)\,G(x-y)\,dy.$$

correlation between values of $u(t)$ at two different times. Since

$$\int_{-\infty}^{\infty} \frac{1}{(n^2-\kappa^2)^2+\kappa^2\lambda^2} e^{i\kappa\tau}\, d\kappa = \frac{\pi}{\lambda n n'} e^{-\frac{1}{2}\lambda|\tau|} \sin\left(n'|\tau| + \sin^{-1}\frac{n'}{n}\right),$$

where $n' = (n^2 - \frac{1}{4}\lambda^2)^{\frac{1}{2}}$ is the damped angular frequency of the body, and since

$$\int_{-\infty}^{\infty} \Phi(\kappa)\, e^{i\kappa\tau}\, d\kappa = \overline{u(t)\,u(t+\tau)} = R(\tau),$$

we have

$$\overline{X(t)\,X(t+\tau)} = \frac{1}{2\lambda n n'} \int_{-\infty}^{\infty} R(\tau-\xi)\, e^{-\frac{1}{2}\lambda|\xi|} \sin\left(n'|\xi| + \sin^{-1}\frac{n'}{n}\right) d\xi. \tag{4.1.6}$$

(4.1.4) and (4.1.6) are the most important of the desired relations between the statistical parameters of $X(t)$ and $u(t)$; similar relations for other statistical parameters can be obtained without difficulty.

Turning now to the physical significance of the problem which we have solved, an immediate application is to the action of an electrical band-pass filter circuit on a random input signal (Burgers, 1948*b*). $\dot{X}(t)$ and $u(t)$ would then represent the output and input signal currents respectively, n the natural (undamped) frequency of the circuit and λ the inductive damping coefficient. The important relation in this case is (4.1.4), which shows how the portion of the input spectrum passed by the filter circuit depends on the characteristics of the circuit. When the damping is sufficiently small the circuit is sharply tuned to the resonant frequency n and a measurement of $\overline{X^2} = \int_{-\infty}^{\infty} \Psi(\kappa)\, d\kappa$ is in effect a measurement of $\Phi(n)$. Thus by adjusting n it is possible to measure the whole of the spectrum function of the input signal $u(t)$. This is the practical method of determination of the spectrum of the turbulent velocity recorded by a hot-wire anemometer in a wind-tunnel stream. It is also used for the analysis of electrical and acoustic 'noise'.

Another application of the mathematical analysis is to the motion of a body subject to a restoring force proportional to its displacement from a position of equilibrium and immersed in a fluid which is in statistically steady turbulent motion. The motion of a pendulum swinging in a turbulent fluid (Lin, 1943), or the motion of a buoyant balloon tethered to a fixed point in a turbulent wind, are examples of this kind. The quantity $u(t)$ is here the hydrodynamic force on the body produced by the turbulence, and if we wish to identify $u(t)$ with the turbulent velocity (apart from a constant factor)—as of course we must do for consistency if $-\lambda\dot{X}$ is intended to represent hydrodynamic damping of the system—we must make further assumptions. In cases in which the fluid has zero mean velocity relative to the body the assumption is that $u-\dot{X}$ is small enough for Stokes's law of viscous resistance to be valid. In cases in which the fluid has a finite steady mean velocity relative to the body, the required condition is that the forces due to the turbulent motion should be a perturbation only (whatever the dependence of force on velocity), that is, that the turbulent velocity $u(t)$ should be small compared with the mean wind speed. In all cases the linear dimensions of the body should be small compared with the length characteristic of the turbulent motion in order that it should be possible to speak of the velocity of the body through the fluid in its immediate neighbourhood. Also it should be noted that $R(\tau)$ cannot be identified with the correlation of the turbulent velocity at two different times at a fixed point, unless it is also assumed that the displacement of the body is always small compared with the length characteristic of the energy-containing eddies of the turbulence.

4.2. Passage of a turbulent stream through wire gauze

In this and the subsequent section of this chapter, the problem under discussion is linearized by the assumption that the field of turbulence responds to some external effect very rapidly—so rapidly that the inertia and viscous forces acting on the fluid produce a negligible change only in the distribution of velocity. The response to the external effect (which must itself be linear, of course) is thus assumed to take place within a time which is small compared with a time representative of the decay of the turbulence (say, the time required for the energy to be reduced by half). In such problems we are virtually concerned with the spatial distribution of velocity only. Two or three problems which have some practical usefulness have been solved on this semi-kinematical basis.

The first to be considered is the determination of the effect of translating a field of homogeneous turbulence at a uniform speed U perpendicularly through a plane sheet of wire gauze. This is a problem which arises from the long-established practical use of wire gauzes to damp out disturbances in a wind-tunnel stream (see Dryden and Schubauer, 1947). Details of the construction of the gauze are unimportant for our purpose, except inasmuch as they affect the values of two parameters to be defined below, and we can regard the gauze as a continuous sheet which introduces no length parameter into the problem (as is approximately true in practice when the distance between neighbouring wires of the gauze is sufficiently small). The history of the turbulent velocity at a point moving with the mean stream is then as follows. Far upstream and far downstream of the gauze the velocity field is assumed to be experiencing no change with time (except translation). While the point is in the neighbourhood of the gauze the velocity changes, as the point moves, in a manner demanded by the presence of the gauze and the local velocity distribution. By the phrase 'in the neighbourhood of the gauze', we mean within a distance from the gauze comparable with a length l characteristic of the energy-containing eddies of the turbulence, since this defines the distance over which turbulent velocities are statistically related. The assumption on which the linearization is based is thus that negligible decay of the turbulence occurs while the field of turbulence is translated through the region of influence of the gauze, that is, while the point moving with the mean flow covers a distance of order l; expressed analytically, we require the condition

$$\frac{1}{\overline{\mathbf{u}^2}}\frac{d\overline{\mathbf{u}^2}}{dt} \ll \frac{U}{l} \tag{4.2.1}$$

to be satisfied.

We shall see later that this condition is satisfied when the turbulence has been generated by forcing the stream through a grid of bars at velocity U, which is by far the most important practical case. For in such cases it is found† that $d\overline{\mathbf{u}^2}/dt$ is of the order of $(\overline{\mathbf{u}^2})^{\frac{3}{2}}/l$ (during the important period of the decay), so that the condition becomes

$$\frac{(\overline{\mathbf{u}^2})^{\frac{1}{2}}}{U} \ll 1. \tag{4.2.2}$$

The value of $(\overline{\mathbf{u}^2})^{\frac{1}{2}}/U$ varies with the time of decay of the turbulence, but is at least as small as 0·1 before the turbulence generated by

† See Appendix to § 6.1.

the grid has become homogeneous, and becomes less thereafter. The turbulence is essentially a small disturbance of the uniform stream U.

The two parameters which are assumed to specify the gauze (so far as its action on the turbulence is concerned) are, first, the resistance coefficient $k(\theta)$ defined by

$$p_1 - p_2 = k(\theta) . \tfrac{1}{2}\rho V^2, \tag{4.2.3}$$

where p_1 and p_2 are the pressures in front of and behind the gauze when a uniform stream of speed V (upstream of the gauze) flows through it at an angle θ to the normal to the plane of the gauze, and, second, the deflexion coefficient $\alpha(\theta)$ defined by

$$\phi = \theta\alpha(\theta), \tag{4.2.4}$$

where θ and ϕ are the angles between the normal to the gauze and the directions (assumed coplanar with the normal) of the uniform stream entering and leaving the gauze. These two coefficients describe the forces exerted by the gauze normal and parallel to its plane respectively. We shall assume that when turbulent fluctuations are superimposed on the uniform stream, the instantaneous pressure drop and deflexion at each point of the wire are given by (4.2.3) and (4.2.4) with local instantaneous values of V and θ inserted on the right sides of these equations. In accordance with our basic assumption (4.2.2), we are concerned with small values of θ only, and since $k(\theta)$ and $\alpha(\theta)$ are necessarily stationary at $\theta = 0$, we shall regard k and α as independent of θ. There is some experimental evidence (Schubauer, Spangenberg and Klebanoff, 1950) to show that the approximate empirical relation

$$\alpha(0) = \frac{1 \cdot 1}{[1 + k(0)]^{\frac{1}{2}}} \tag{4.2.5}$$

holds for gauzes of different construction and streams of different speeds; this will simplify our conclusions about the effect of an arbitrary gauze on the turbulence.

We shall represent the turbulent velocity far upstream of the gauze by $\mathbf{u}'$ and that far downstream by $\mathbf{u}''$. If the origin of the spatial coordinate system is in the plane of the gauze, with the positive x_1-axis directed downstream, $\mathbf{u}'$ and $\mathbf{u}''$ are functions of $x_1 - Ut$, x_2 and x_3, where U is the velocity of the stream carrying the turbulence. The transition from the velocity distribution $\mathbf{u}'$ to the distribution $\mathbf{u}''$ takes place wholly in the neighbourhood of the gauze, partly upstream from the gauze, partly downstream and partly as a discontinuous change across the plane of the gauze. Our basic assumption is that the intensity of the turbulence is too small to permit inertia and viscous forces to change the velocity distribution while it is being transported through the region of influence of the gauze; consequently the modification to the velocity distribution which takes place in the neighbourhood of the gauze (except across the gauze itself) must be due entirely to the distribution of pressure drop across the gauze and must therefore be irrotational. Hence, on the upstream side of the gauze, we have

$$\mathbf{u}(\mathbf{x},t)=\mathbf{u}'(x_1-Ut,x_2,x_3)+\nabla\phi'(\mathbf{x},t), \tag{4.2.6}$$

where ϕ' is a potential function which approaches a constant value as $x_1\rightarrow-\infty$, and on the downstream side

$$\mathbf{u}(\mathbf{x},t)=\mathbf{u}''(x_1-Ut,x_2,x_3)+\nabla\phi''(\mathbf{x},t), \tag{4.2.7}$$

where ϕ'' approaches a constant value as $x_1\rightarrow\infty$. ϕ' and ϕ'' are determined by the pressure-drop equation (4.2.3), which provides the necessary boundary condition at the plane $x_1=0$. For, consistent with our basic assumption, the equations of motion near the gauze on the upstream and downstream sides are

$$\rho\left(\frac{\partial}{\partial t}+U\frac{\partial}{\partial x_1}\right)\phi'=-p \tag{4.2.8}$$

and

$$\rho\left(\frac{\partial}{\partial t}+U\frac{\partial}{\partial x_1}\right)\phi''=-p-k.\tfrac{1}{2}\rho U^2 \tag{4.2.9}$$

(the term $-k.\frac{1}{2}\rho U^2$ represents the level of the pressure at $x=+\infty$), and if these equations are taken at $x_1=-0$ and $x_1=+0$ respectively and are subtracted, we have as the joint boundary condition on ϕ' and ϕ''

$$\begin{aligned}\left[\left(\frac{\partial}{\partial t}+U\frac{\partial}{\partial x_1}\right)(\phi''-\phi')\right]_{x_1=0}&=\frac{(p)_{-0}-(p)_{+0}}{\rho}-k.\tfrac{1}{2}U^2\\&=k.\tfrac{1}{2}[(U+u_1)^2+u_2^2+u_3^2-U^2]_{x_1=-0}\\&\approx kU(u_1)_{x_1=0}=kU\left(u_1'+\frac{\partial\phi'}{\partial x_1}\right)_{x_1=0},\end{aligned} \tag{4.2.10}$$

where the suffix 1 denotes a component normal to the gauze.

The modification to the velocity distribution across the plane of the gauze is seen from (4.2.4) to be given by

$$(u_2)_{x_1=+0}=\alpha(u_2)_{x_1=-0}, \quad (u_3)_{x_1=+0}=\alpha(u_3)_{x_1=-0}. \qquad (4.2.11)$$

The conservation of mass prohibits any discontinuity in the component u_1, so that

$$(u_1)_{x_1=+0}=(u_1)_{x_1=-0}. \qquad (4.2.12)$$

These conditions are sufficient to relate the statistical characteristics of $\mathbf{u}'$ and $\mathbf{u}''$ and so to determine the effect of the gauze on the turbulence passing through it. We make a Fourier resolution of the velocity field, in the manner of (2.5.3), and, in virtue of the linearity of the above equations, may consider the effect of the gauze on a single Fourier component in isolation. We write

$$\begin{Bmatrix}\mathbf{u}'\\ \mathbf{u}''\end{Bmatrix}(x_1-Ut, x_2, x_3)=\int e^{\iota(\boldsymbol{\kappa}\cdot\mathbf{x}-\kappa_1 Ut)}\, d\begin{Bmatrix}\mathbf{Z}'\\ \mathbf{Z}''\end{Bmatrix}(\boldsymbol{\kappa}), \qquad (4.2.13)$$

and, since ϕ' and ϕ'' are stationary random functions of x_2 and x_3,

$$\begin{Bmatrix}\phi'\\ \phi''\end{Bmatrix}(\mathbf{x}, t)=\int e^{\iota(\kappa_2 x_2+\kappa_3 x_3)}\, d\begin{Bmatrix}\Lambda'\\ \Lambda''\end{Bmatrix}(\kappa_2, \kappa_3, x_1, t). \qquad (4.2.14)$$

Now ϕ' and ϕ'' are potential functions, so that

$$\left(\frac{\partial^2}{\partial x_1^2}-\tau^2\right) d\begin{Bmatrix}\Lambda'\\ \Lambda''\end{Bmatrix}=0,$$

where $\tau^2=\kappa_2^2+\kappa_3^2$, and

$$d\Lambda'(\kappa_2, \kappa_3, x_1, t) \propto e^{\tau x_1}, \quad d\Lambda''(\kappa_2, \kappa_3, x_1, t) \propto e^{-\tau x_1}. \qquad (4.2.15)$$

Moreover, ϕ' and ϕ'' are stationary random functions of t. Hence we can express ϕ' and ϕ'' in terms of new Fourier coefficients $d\Gamma'(\boldsymbol{\kappa})$ and $d\Gamma''(\boldsymbol{\kappa})$:

$$\left.\begin{aligned}\phi'(\mathbf{x}, t)&=\int e^{\tau x_1+\iota(\boldsymbol{\tau}\cdot\mathbf{x}-\kappa_1 Ut)}\, d\Gamma'(\boldsymbol{\kappa}),\\ \phi''(\mathbf{x}, t)&=\int e^{-\tau x_1+\iota(\boldsymbol{\tau}\cdot\mathbf{x}-\kappa_1 Ut)}\, d\Gamma''(\boldsymbol{\kappa}),\end{aligned}\right\} \qquad (4.2.16)$$

where $\boldsymbol{\tau}$ is the vector component of $\boldsymbol{\kappa}$ in the plane of the gauze.

Substituting these integrals in the equations (4.2.10), (4.2.11) and (4.2.12), we find for the general Fourier coefficient the relations

$$(-\iota\kappa_1 - \tau)\,\mathrm{d}\Gamma'' - (-\iota\kappa_1 + \tau)\,\mathrm{d}\Gamma' = k(\mathrm{d}Z_1' + \tau\,\mathrm{d}\Gamma'), \quad (4.2.17)$$

$$\mathrm{d}Z_2'' + \iota\kappa_2\,\mathrm{d}\Gamma'' = \alpha(\mathrm{d}Z_2' + \iota\kappa_2\,\mathrm{d}\Gamma'), \quad (4.2.18)$$

$$\mathrm{d}Z_3'' + \iota\kappa_3\,\mathrm{d}\Gamma'' = \alpha(\mathrm{d}Z_3' + \iota\kappa_3\,\mathrm{d}\Gamma'), \quad (4.2.19)$$

$$\mathrm{d}Z_1'' - \tau\,\mathrm{d}\Gamma'' = \mathrm{d}Z_1' + \tau\,\mathrm{d}\Gamma', \quad (4.2.20)$$

respectively. Together with the continuity condition

$$\boldsymbol{\kappa}.\mathrm{d}\mathbf{Z}' = \boldsymbol{\kappa}.\mathrm{d}\mathbf{Z}'' = 0, \quad (4.2.21)$$

the above equations are sufficient to determine the five unknown quantities $\mathrm{d}\mathbf{Z}''$, $\mathrm{d}\Gamma'$ and $\mathrm{d}\Gamma''$, in terms of $\mathrm{d}\mathbf{Z}'$ and the parameters k and α. The solution for $\mathrm{d}\mathbf{Z}''$ is readily found to be

$$\mathrm{d}Z_1''(\boldsymbol{\kappa}) = \frac{(\beta+\iota)\,[2\alpha\beta + \iota(\alpha k - 1 - \alpha)]}{(\beta-\iota)\,[2\beta + \iota(k+1+\alpha)]}\,\mathrm{d}Z_1'(\boldsymbol{\kappa}) = J(\beta)\,.\,\mathrm{d}Z_1'(\boldsymbol{\kappa}), \quad \text{say}, \quad (4.2.22)$$

$$\mathrm{d}Z_2''(\boldsymbol{\kappa}) = \alpha\,\mathrm{d}Z_2'(\boldsymbol{\kappa}) + \frac{\kappa_1\kappa_2}{\kappa_2^2+\kappa_3^2}\,[\alpha - J(\beta)]\,\mathrm{d}Z_1'(\boldsymbol{\kappa}), \quad (4.2.23)$$

$$\mathrm{d}Z_3''(\boldsymbol{\kappa}) = \alpha\,\mathrm{d}Z_3'(\boldsymbol{\kappa}) + \frac{\kappa_1\kappa_3}{\kappa_2^2+\kappa_3^2}\,[\alpha - J(\beta)]\,\mathrm{d}Z_1'(\boldsymbol{\kappa}), \quad (4.2.24)$$

where $\beta = \kappa_1/\tau = \kappa_1/(\kappa_2^2+\kappa_3^2)^{\frac{1}{2}}$ is a measure of the angle which the wave-number vector makes with the normal to the plane of the gauze. The solution may also be written concisely as

$$\mathrm{d}Z_i''(\boldsymbol{\kappa}) = \alpha\,\mathrm{d}Z_i'(\boldsymbol{\kappa}) + \left(\frac{\kappa^2\delta_{1i} - \kappa_1\kappa_i}{\kappa_2^2+\kappa_3^2}\right)[J(\beta) - \alpha]\,\mathrm{d}Z_1'(\boldsymbol{\kappa}), \quad (4.2.25)$$

from which the effect of the gauze on any statistical characteristics of the turbulence may be evaluated.

There is great practical interest in the amount of turbulent energy suppressed by the wire gauze, so that we consider the relation between the spectrum tensors of the turbulence far upstream and far downstream from the gauze. The relation between the spectrum function $\Phi_{ij}(\boldsymbol{\kappa})$ and the increment $dZ_i(\boldsymbol{\kappa})$ is given by (2.5.5), and, with an obvious notation, (4.2.25) yields the result

$$\Phi''_{ij}(\boldsymbol{\kappa}) = \alpha^2\Phi'_{ij}(\boldsymbol{\kappa}) + \frac{(\kappa^2\delta_{1i}-\kappa_1\kappa_i)(\kappa^2\delta_{1j}-\kappa_1\kappa_j)}{(\kappa_2^2+\kappa_3^2)^2}(JJ^*-\alpha^2)\Phi'_{11}(\boldsymbol{\kappa})$$

$$+\alpha\left(\frac{\kappa^2\delta_{1i}-\kappa_1\kappa_i}{\kappa_2^2+\kappa_3^2}\right)(J^*-\alpha)\left(\Phi'_{1j}-\frac{\kappa^2\delta_{1j}-\kappa_1\kappa_j}{\kappa_2^2+\kappa_3^2}\Phi'_{11}\right)$$

$$+\alpha\left(\frac{\kappa^2\delta_{1j}-\kappa_1\kappa_j}{\kappa_2^2+\kappa_3^2}\right)(J-\alpha)\left(\Phi'_{i1}-\frac{\kappa^2\delta_{i1}-\kappa_1\kappa_i}{\kappa_2^2+\kappa_3^2}\Phi'_{11}\right). \quad (4.2.26)$$

In particular, we are interested in the reduction in the contributions to the energy of the velocity components normal and parallel to the gauze. Far downstream the spectrum densities, at wave-number $\boldsymbol{\kappa}$, of the energy in the longitudinal and lateral motions are

$$\Phi''_{11}(\boldsymbol{\kappa}) = JJ^*\Phi'_{11}(\boldsymbol{\kappa}), \quad (4.2.27)$$

$$\Phi''_{22}(\boldsymbol{\kappa})+\Phi''_{33}(\boldsymbol{\kappa}) = \alpha^2(\Phi'_{22}+\Phi'_{33})+\beta^2(JJ^*-\alpha^2)\Phi'_{11}, \quad (4.2.28)$$

respectively, and consequently the factors giving the reduction in the total energy of the longitudinal and lateral velocity components are

$$\mu = \frac{\overline{u_1''^2}}{\overline{u_1'^2}} = \frac{\int JJ^*\Phi'_{11}(\boldsymbol{\kappa})\,d\boldsymbol{\kappa}}{\int \Phi'_{11}(\boldsymbol{\kappa})\,d\boldsymbol{\kappa}}, \quad (4.2.29)$$

$$\nu = \frac{\overline{u_2''^2}+\overline{u_3''^2}}{\overline{u_2'^2}+\overline{u_3'^2}} = \alpha^2 + \frac{\int \beta^2(JJ^*-\alpha^2)\Phi'_{11}(\boldsymbol{\kappa})\,d\boldsymbol{\kappa}}{\int (\Phi'_{22}+\Phi'_{33})\,d\boldsymbol{\kappa}}. \quad (4.2.30)$$

From the definition of $J(\beta)$ (see (4.2.22)) we have

$$\beta^2(JJ^*-\alpha^2) = \tfrac{1}{4}(\alpha k-1-\alpha)^2 - \tfrac{1}{4}(k+1+\alpha)^2 JJ^*,$$

so that $$\nu = \alpha^2 + \tfrac{1}{4}[(\alpha k-1-\alpha)^2-(k+1+\alpha)^2\mu]\frac{\overline{u_1'^2}}{\overline{u_2'^2}+\overline{u_3'^2}}, \quad (4.2.31)$$

giving ν as a function of μ and the directional distribution of energy in the turbulence far upstream. The integrals in (4.2.29) can be evaluated when the spectrum of the turbulence approaching the gauze is known.

The special case in which the turbulence far upstream is isotropic is important, and the integrals are then very simple. We have seen that in this case

$$\Phi'_{ij}(\mathbf{x}) = \frac{E(\kappa)}{4\pi\kappa^4}(\kappa^2\delta_{ij} - \kappa_i\kappa_j),$$

and the reduction factor (4.2.29) becomes

$$\mu = \frac{\displaystyle\int \frac{E(\kappa)}{\kappa^2(1+\beta^2)} JJ^* \mathrm{d}\mathbf{x}}{\displaystyle\int \frac{E(\kappa)}{\kappa^2(1+\beta^2)} \mathrm{d}\mathbf{x}}.$$

J is a function of β $(=\kappa_1/(\kappa_2^2+\kappa_3^2)^{\frac{1}{2}})$ only, so that if the volume integration is carried out in terms of spherical polar coordinates, the integration with respect to κ cancels, giving a value for μ which is independent of the form of the function $E(\kappa)$. We find

$$\mu = \frac{\displaystyle\int_{-\infty}^{\infty} (1+\beta^2)^{-\frac{3}{2}} JJ^* \mathrm{d}\beta}{\displaystyle\int_{-\infty}^{\infty} (1+\beta^2)^{-\frac{3}{2}} \mathrm{d}\beta}$$

$$= \frac{3}{2}\int_0^{\infty} (1+\beta^2)^{-\frac{3}{2}} \frac{4\alpha^2\beta^2 + (\alpha k - 1 - \alpha)^2}{4\beta^2 + (k+1+\alpha)^2} \mathrm{d}\beta, \qquad (4.2.32)$$

and from (4.2.31) the reduction in lateral energy is

$$\nu = \alpha^2 + \tfrac{1}{4}(\alpha k - 1 - \alpha)^2 - \tfrac{1}{8}\mu(k+1+\alpha)^2. \qquad (4.2.33)$$

The difference in the values of μ and ν is a reflexion of the obvious result that the turbulence is not isotropic after passing through the gauze, and is axially symmetrical about the normal to the gauze. Numerical values of the factors μ and ν for various values of k are shown in fig. 4.1, where it has been assumed that the relation (4.2.5) between α and k is valid.

The effect of the gauze on other parameters of the turbulence can be calculated. An interesting result is obtained by putting $\kappa_1 = 0$ in the expressions (4.2.22), (4.2.23) and (4.2.24) for the effect of the gauze on the general Fourier coefficient, viz.

$$\mathrm{d}Z''_1(0, \kappa_2, \kappa_3) = \frac{1+\alpha-\alpha k}{1+\alpha+k}\mathrm{d}Z'_1(0, \kappa_2, \kappa_3), \qquad (4.2.34)$$

$$\mathrm{d}\begin{Bmatrix} Z''_2 \\ Z''_3 \end{Bmatrix}(0, \kappa_2, \kappa_3) = \alpha \mathrm{d}\begin{Bmatrix} Z'_2 \\ Z'_3 \end{Bmatrix}(0, \kappa_2, \kappa_3). \qquad (4.2.35)$$

These equations describe the effect of the gauze on a disturbance which, far upstream, has no variation in the x_1-direction, that is, one which, in isolation, is 'steady' in the sense that there is no change in the velocity at a fixed point as the disturbance field is carried downstream. The effects of the gauze on the longitudinal and lateral velocity components are here independent, the reduction in

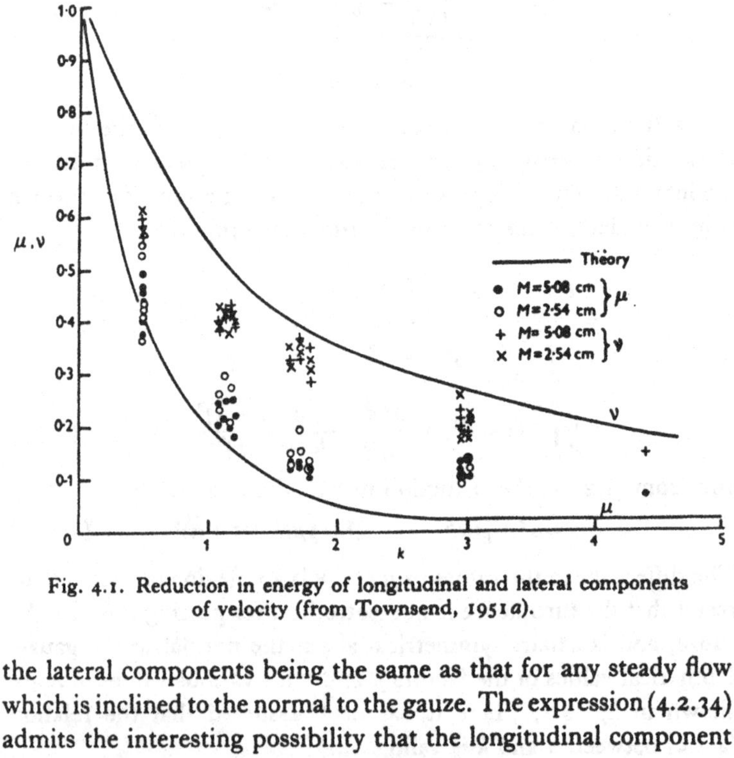

Fig. 4.1. Reduction in energy of longitudinal and lateral components of velocity (from Townsend, 1951*a*).

the lateral components being the same as that for any steady flow which is inclined to the normal to the gauze. The expression (4.2.34) admits the interesting possibility that the longitudinal component is entirely suppressed by a gauze such that

$$k = 1 + \alpha^{-1},$$

which corresponds to $k \approx 2{\cdot}8$ if α and k are related as in (4.2.5). For larger values of k the longitudinal velocity is reversed in sign and also reduced in magnitude. The longitudinal component of the velocity for other wave-numbers is not suppressed when $k = 2{\cdot}8$, but nevertheless it will be noticed from fig. 4.1 that μ has a shallow minimum for a value of k between 3 and 4.

The above theory of the effect of wire gauze has been submitted to experimental test and its approximate general validity has been established. For details of some measurements of the reduction of a steady disturbance, the paper by Taylor and Batchelor (1949), in which the theory was first presented, may be consulted; further measurements of the same kind have been described by Schubauer,

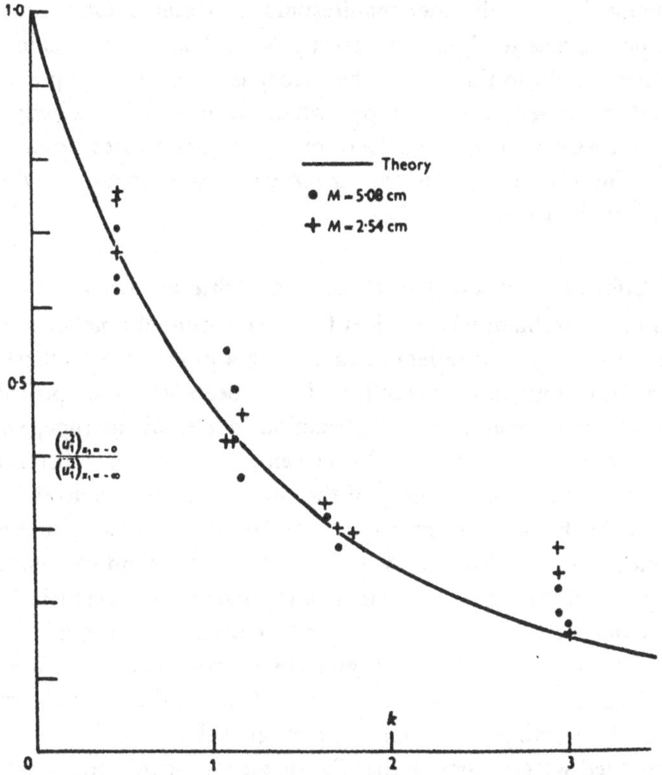

Fig. 4.2. Reduction in energy of longitudinal component immediately upstream from gauze (from Townsend, 1951 *a*).

Spangenberg and Klebanoff (1950). Townsend (1951 *a*) has recently published an account of extensive measurements of the effect of gauze on isotropic turbulence, and his measured values of μ and ν are reproduced in fig. 4.1. The principal defect of the theory, as revealed by comparison with the measurements shown in this figure, is that it predicts too great a degree of anisotropy downstream from the gauze. Townsend has shown that the reduction in total energy (i.e. $\mu + 2\nu$) is predicted quite accurately by the theory (as can be seen from fig. 4.1), and he has also established that the value of $\overline{u_1^2}$ immediately upstream from the gauze (i.e. at $x_1 = -0$) is given to a good approximation by the theory, as shown in fig. 4.2. It appears that in the neighbourhood of the gauze, on the downstream side, there is a strong tendency to isotropy which leaves the total energy unchanged. Like all other manifestations of inertia forces in the turbulence, this tendency to isotropy is not taken into account in the theory, and to this extent the theory is in error. It is possible that this tendency to isotropy, which seems to act chiefly in the downstream neighbourhood of the gauze, arises from the discontinuous change in the triple-velocity correlation which occurs at the gauze.

4.3. Effect of sudden distortion of a turbulent stream

Another problem which arises from wind-tunnel practice is the effect of passing a turbulent stream along a pipe in which there is a sudden change in cross-section. It has been known empirically for many years that a sharp contraction in area of the pipe, with consequent acceleration of the stream, produces a substantial decrease in the kinetic energy of the turbulence relative to that of the stream. By placing very large contractions, or throats, at a short distance upstream from the working section of a wind tunnel, the designer is able to obtain a stream from which undesirable turbulent fluctuations have been almost wholly eliminated. A qualitative explanation of the effect was first given by Prandtl† from a consideration of the effect on steady disturbances (that is, those which have no variation with position in the direction of the stream).

Provided we can assume that the distortion of the stream takes place so rapidly that inertia and viscous forces arising from the turbulent motion have no effect, the problem becomes linear, as was shown first by G. I. Taylor (1935 *b*). On this basis the problem can be solved in much the same way as that considered in the previous section. We imagine that in the regions upstream and downstream of the distorting section of the pipe the mean flow is uniform and the turbulence is homogeneous. The way in which

† L. Prandtl, *Handbuch der Experimentalphysik*, Leipzig, 1932, 4, part 2, 73.

the change in the turbulence occurs during the process of distortion is irrelevant to the linearized theory; only the initial and final states concern us. The condition on which turbulent inertia and viscous forces can be ignored during the process of distortion of the stream is

$$\frac{1}{\overline{\mathbf{u}^2}}\frac{d\overline{\mathbf{u}^2}}{dt} \ll \frac{U}{D}, \tag{4.3.1}$$

where U is the speed of the stream and D is the length, in the stream direction, of the distorting section of the pipe. Making use again of the known result that $d\overline{\mathbf{u}^2}/dt$ is of the order of $(\overline{\mathbf{u}^2})^{\frac{3}{2}}/l$, where l is the length scale of the turbulence, we require

$$\frac{(\overline{\mathbf{u}^2})^{\frac{1}{2}}}{U} \ll \frac{l}{D}. \tag{4.3.2}$$

$(\overline{\mathbf{u}^2})^{\frac{1}{2}}$ is always small compared with U, but l/D is usually small also, and cases in which (4.3.2) is satisfied will be the exception rather than the rule. The results obtained below are therefore less useful than those of the preceding section.

The linearized problem may be solved by considering the way in which the vorticity of the turbulent motion is changed and redistributed by the distortion of the stream. To this end we use a Lagrangian specification of the motion, and let $\mathbf{u}'(\mathbf{a})$ and $\boldsymbol{\omega}'(\mathbf{a})$ be the velocity and vorticity associated with the fluid element which has position $\mathbf{a}$ in the field of turbulence upstream of the distortion, relative to axes moving with the mean velocity. In the field of turbulence downstream of the distortion, this same fluid element has position $\mathbf{x}$ and the corresponding velocity and vorticity are $\mathbf{u}''(\mathbf{x})$ and $\boldsymbol{\omega}''(\mathbf{x})$, relative to axes moving with the new mean velocity. In the absence of viscous forces the final vorticity $\boldsymbol{\omega}''(\mathbf{x})$ is determined entirely by the change in relative position of the fluid particles, and is given by Cauchy's equation†

$$\omega_i''(\mathbf{x}) = \frac{\partial x_i}{\partial a_j}\omega_j'(\mathbf{a}), \tag{4.3.3}$$

that is,

$$\epsilon_{ipq}\frac{\partial u_q''(\mathbf{x})}{\partial x_p} = \epsilon_{jpq}\frac{\partial x_i}{\partial a_j}\frac{\partial u_q'(\mathbf{a})}{\partial a_p} \tag{4.3.4}$$

† See H. Lamb, *Hydrodynamics*, Oxford University Press, 6th ed., chap. 7.

We now make the assumption that $\partial x_i/\partial a_j$ is a constant, independent of $\mathbf{a}$. This assumption implies, in accordance with our basic hypothesis, that the strain experienced by any fluid element during the distortion process does not depend on the turbulent motion. It also contains the approximation that the contribution to the strain from the passage of the fluid element through the distribution of mean velocity in the distorting section is the same for all fluid particles; this is certainly not exactly true, in general, but it will be a sufficiently good approximation when the scale of the turbulence is small compared with the tunnel width. The strain imposed by the distortion and described by the tensor $\partial x_i/\partial a_j$ is thus homogeneous. On this basis, the operation of taking the curl (with respect of $\mathbf{x}$) of both sides of (4.3.4) gives

$$\epsilon_{rst}\epsilon_{ipq}\frac{\partial^2 u_q''(\mathbf{x})}{\partial x_p\,\partial x_s}=\epsilon_{rst}\epsilon_{jpq}\frac{\partial x_i}{\partial a_j}\frac{\partial}{\partial x_s}\left[\frac{\partial u_q'(\mathbf{a})}{\partial a_p}\right];$$

but

$$\epsilon_{irs}\epsilon_{ipq}=\delta_{rp}\delta_{sq}-\delta_{rq}\delta_{sp}, \tag{4.3.5}$$

and hence

$$-\nabla_x^2 u_r''(\mathbf{x})=\epsilon_{rst}\epsilon_{jpq}\frac{\partial x_i}{\partial a_j}\frac{\partial a_t}{\partial x_s}\frac{\partial^2 u_q'(\mathbf{a})}{\partial a_p\,\partial a_t}. \tag{4.3.6}$$

The statistical characteristics of the turbulence fields existing before and after the distortion may be related with the aid of this equation. We make Fourier resolutions of the velocity distributions:

$$\mathbf{u}''(\mathbf{x})=\int e^{i\boldsymbol{\chi}.\mathbf{x}}\,d\mathbf{Z}''(\boldsymbol{\chi}),\quad \mathbf{u}'(\mathbf{a})=\int e^{i\boldsymbol{\kappa}.\mathbf{a}}\,d\mathbf{Z}'(\boldsymbol{\kappa}), \tag{4.3.7}$$

and find from (4.3.6) that the components of $d\mathbf{Z}''(\boldsymbol{\chi})$, where $\boldsymbol{\chi}$ is the wave-number vector such that

$$\boldsymbol{\chi}.\mathbf{x}=\boldsymbol{\kappa}.\mathbf{a},\quad \text{i.e.}\quad \chi_i=\kappa_j\frac{\partial a_j}{\partial x_i}, \tag{4.3.8}$$

are related to those of $d\mathbf{Z}'(\boldsymbol{\kappa})$ by the equation

$$\chi^2 dZ_r''(\boldsymbol{\chi})=-\epsilon_{rst}\epsilon_{jpq}\frac{\partial x_i}{\partial a_j}\frac{\partial a_t}{\partial x_s}\kappa_p\kappa_t\,dZ_q'(\boldsymbol{\kappa}). \tag{4.3.9}$$

As in the problem of the preceding section, practical interest is concentrated chiefly on the change in the spectrum tensor of the turbulence. The expression for $\Phi_{ij}''(\boldsymbol{\chi})$ in terms of $\Phi_{pq}'(\boldsymbol{\kappa})$ follows immediately from (4.3.9), and is given by

$$\Phi''_{rl}(\chi)\,d\chi = \frac{\epsilon_{rsi}\epsilon_{jpq}\dfrac{\partial x_i}{\partial a_j}\dfrac{\partial a_t}{\partial x_s}\epsilon_{lmu}\epsilon_{vab}\dfrac{\partial x_u}{\partial a_v}\dfrac{\partial a_n}{\partial x_m}}{\left(\kappa_c\kappa_d\dfrac{\partial a_c}{\partial x_e}\dfrac{\partial a_d}{\partial x_e}\right)^2}\kappa_p\kappa_t\kappa_a\kappa_n\Phi'_{qb}(\boldsymbol{\kappa})\,d\boldsymbol{\kappa}. \tag{4.3.10}$$

The case in which $\partial x_i/\partial a_j$ represents a pure strain is almost the only one of any importance for wind-tunnel practice, and the analysis will be restricted accordingly. With the principal axes of the pure strain as axes of reference, and with e_1-1, e_2-1, e_3-1 as the principal extensions, the components of the strain tensor become

$$\frac{\partial x_1}{\partial a_1}=e_1,\quad \frac{\partial x_2}{\partial a_2}=e_2,\quad \frac{\partial x_3}{\partial a_3}=e_3,\quad \frac{\partial x_i}{\partial a_j}=0 \quad \text{if} \quad i\neq j,$$

while those of the reciprocal strain tensor are

$$\frac{\partial a_1}{\partial x_1}=\frac{1}{e_1},\quad \frac{\partial a_2}{\partial x_2}=\frac{1}{e_2},\quad \frac{\partial a_3}{\partial x_3}=\frac{1}{e_3},\quad \frac{\partial a_i}{\partial x_j}=0 \quad \text{if} \quad i\neq j,$$

where $e_1e_2e_3=1$ in order to satisfy the continuity equation. Equation (4.3.9) now becomes

$$dZ''_1(\chi)=\frac{1}{e_1}dZ'_1(\boldsymbol{\kappa})-\frac{\kappa_1}{e_1\chi^2}\left[\frac{\kappa_1}{e_1^2}dZ'_1(\boldsymbol{\kappa})+\frac{\kappa_2}{e_2^2}dZ'_2(\boldsymbol{\kappa})+\frac{\kappa_3}{e_3^2}dZ'_3(\boldsymbol{\kappa})\right] \tag{4.3.11}$$

with two similar equations, and the diagonal components (which, in view of the continuity equation, are sufficient to determine all components) of $\Phi''_{ij}(\chi)$ are given by

$$\begin{aligned}\Phi''_{11}(\chi)=\frac{1}{e_1^2}\Phi'_{11}(\boldsymbol{\kappa})&+\frac{\kappa_1^2}{e_1^2\chi^4}\left[\left(\frac{\kappa_2^2}{e_2^2}+\frac{\kappa_3^2}{e_3^2}\right)\left(\frac{1}{e_2^2}+\frac{1}{e_3^2}-\frac{2}{e_1^2}\right)+\kappa_1^2\left(e_1^2-\frac{1}{e_1^4}\right)\right]\Phi'_{11}(\boldsymbol{\kappa})\\&+\frac{\kappa_2^2}{e_2^2\chi^4}(e_3^2-e_1^2)\left(\kappa_1^2+\kappa_2^2+\frac{e_2^2}{e_3^2}\kappa_3^2\right)\Phi'_{22}(\boldsymbol{\kappa})\\&+\frac{\kappa_3^2}{e_3^2\chi^4}(e_2^2-e_1^2)\left(\kappa_1^2+\frac{e_3^2}{e_2^2}\kappa_2^2+\kappa_3^2\right)\Phi'_{33}(\boldsymbol{\kappa}),\end{aligned} \tag{4.3.12}$$

with two similar equations, where the components of χ are κ_1/e_1, κ_2/e_2 and κ_3/e_3.

When the turbulence is isotropic before distortion occurs, we have

$$\Phi'_{11}(\boldsymbol{\kappa}) = \frac{E(\kappa)}{4\pi\kappa^4}(\kappa_2^2+\kappa_3^2), \quad \Phi'_{22}(\boldsymbol{\kappa}) = \text{etc.},$$

and in this case (4.3.12) becomes, after some reduction,

$$\Phi''_{11}(\boldsymbol{\chi}) = \frac{E(\kappa)}{4\pi e_1^2 \chi^4 \kappa^2}\left[\kappa_1^2\left(\frac{\kappa_2^2}{e_2^4}+\frac{\kappa_3^2}{e_3^4}\right)+\left(\frac{\kappa_2^2}{e_2^2}+\frac{\kappa_3^2}{e_3^2}\right)^2\right]. \quad (4.3.13)$$

Likewise the spectrum of total energy in this case is

$$\Phi''_{ii}(\boldsymbol{\chi}) = \frac{E(\kappa)}{4\pi\chi^2\kappa^2}\left[\kappa_1^2(e_2^2+e_3^2)+\kappa_2^2(e_3^2+e_1^2)+\kappa_3^2(e_1^2+e_2^2)\right]. \quad (4.3.14)$$

The reduction in the energy of each velocity component owing to the distortion can now be found by integrating over all wave-numbers.

The formulae for general values of e_1, e_2 and e_3 involve complicated integrals, but become simple if $e_1 = c$, $e_2 = e_3 = 1/\sqrt{c}$; this case has some practical importance since it represents the effect of a symmetrical contraction, which is usually present in wind tunnels of circular or square section. Equation (4.3.13) and the sum of its two partners then reduce to

$$\Phi''_{11}(\boldsymbol{\chi}) = \frac{E(\kappa)}{4\pi\chi^4}(\kappa_2^2+\kappa_3^2), \quad (4.3.15)$$

$$\Phi''_{22}(\boldsymbol{\chi})+\Phi''_{33}(\boldsymbol{\chi}) = \frac{E(\kappa)}{4\pi\kappa^2\chi^4}\left(c\chi^4+\frac{\kappa_1^2\kappa^2}{c^3}\right), \quad (4.3.16)$$

where $\chi^2 = \kappa_1^2/c^2 + c(\kappa_2^2+\kappa_3^2)$. The ratio of the values of the energy of the longitudinal velocity component (i.e. along the axis of symmetry) after and before the distortion is

$$\mu = \frac{\overline{u_1''^2}}{\overline{u_1'^2}} = \frac{\int \Phi''_{11}(\boldsymbol{\chi})\,d\boldsymbol{\chi}}{\int \Phi'_{11}(\boldsymbol{\kappa})\,d\boldsymbol{\kappa}} = \frac{\int E(\kappa)\dfrac{\kappa_2^2+\kappa_3^2}{\chi^4}\,d\boldsymbol{\kappa}}{\int E(\kappa)\dfrac{\kappa_2^2+\kappa_3^2}{\kappa^4}\,d\boldsymbol{\kappa}}.$$

In terms of polar coordinates κ, θ, where

$$\kappa_1 = \kappa\cos\theta, \quad (\kappa_2^2+\kappa_3^2)^{\frac{1}{2}} = \kappa\sin\theta,$$

$$\mu = \frac{\displaystyle\int_0^\pi \frac{\sin^3\theta}{(c^{-2}\cos^2\theta + c\sin^2\theta)^2}\,d\theta}{\displaystyle\int_0^\pi \sin^3\theta\,d\theta}$$

$$= \frac{3}{4c^2}\left[\frac{1+\alpha^2}{2\alpha^3}\log\frac{1+\alpha}{1-\alpha} - \alpha^{-2}\right], \tag{4.3.17}$$

where $\alpha^2 = 1 - c^{-3}$. The ratio of the values of the energy of the lateral velocity components after and before distortion is

$$\nu = \frac{\overline{u_2''^2}+\overline{u_3''^2}}{\overline{u_2'^2}+\overline{u_3'^2}} = \frac{\int[\Phi_{22}''(\chi)+\Phi_{33}''(\chi)]\,d\chi}{\int[\Phi_{22}'(\varkappa)+\Phi_{33}'(\varkappa)]\,d\varkappa}$$

$$= \frac{\displaystyle\int E(\kappa)\left(\frac{c^4\chi^4+\kappa_1^2\kappa^2}{c^3\kappa^2\chi^4}\right)d\varkappa}{\displaystyle\int E(\kappa)\left(\frac{\kappa^2+\kappa_1^2}{\kappa^4}\right)d\varkappa}$$

$$= \frac{c\displaystyle\int_0^\pi [1+c^{-4}\cos^2\theta\,(c^{-2}\cos^2\theta + c\sin^2\theta)^{-2}]\sin\theta\,d\theta}{\displaystyle\int_0^\pi [1+\cos^2\theta]\sin\theta\,d\theta}$$

$$= \frac{3c}{4} + \frac{3}{4}c^{-2}\left[\frac{1}{2\alpha^2} - \frac{1-\alpha^2}{4\alpha^3}\log\frac{1+\alpha}{1-\alpha}\right]. \tag{4.3.18}$$

The ratio of the values of the total kinetic energy of the turbulence before and after distortion is, from (4.3.17) and (4.3.18),

$$\frac{\overline{u_i''u_i''}}{\overline{u_i'u_i'}} = \tfrac{1}{3}(\mu+2\nu) = \frac{c}{2} + \frac{1}{4c^2\alpha}\log\frac{1+\alpha}{1-\alpha}$$

$$= \frac{c}{2} + \frac{1}{2c^2}(1-c^{-3})^{-\frac{1}{2}}\log c^{\frac{3}{2}}\{1+(1-c^{-3})^{\frac{1}{2}}\}. \tag{4.3.19}$$

Numerical values of μ and ν for a symmetrical contraction of isotropic turbulence, as given by (4.3.17) and (4.3.18), are shown in fig. 4.3. It will be noticed that the energy of the lateral components increases (for $c > 1$), while that of the longitudinal component decreases, as expected from Prandtl's original argument that vortices parallel to the x_1-axis are extended and strengthened while vortices directed across the axis of symmetry are weakened. The change in the lateral energy dominates the total energy of the turbulence, which is always increased by the contraction of the stream. However, the energy of the turbulence relative to that of the stream transporting it—which is the significant quantity so far as the effect of turbulence as an aerodynamic disturbance is concerned—is altered by the factors μ/c^2 and ν/c^2, and hence both longitudinal and lateral relative energies are considerably reduced.

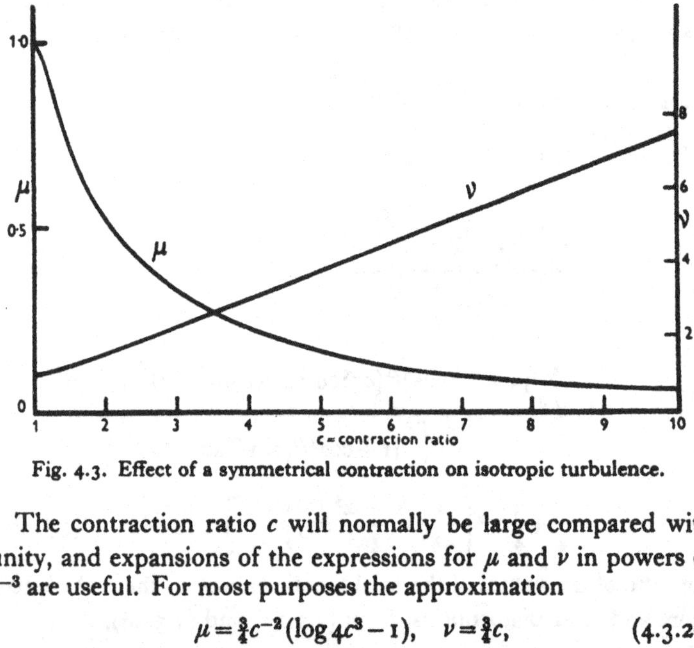

Fig. 4.3. Effect of a symmetrical contraction on isotropic turbulence.

The contraction ratio c will normally be large compared with unity, and expansions of the expressions for μ and ν in powers of c^{-3} are useful. For most purposes the approximation

$$\mu = \tfrac{3}{4}c^{-2}(\log 4c^3 - 1), \quad \nu = \tfrac{3}{4}c, \tag{4.3.20}$$

will be sufficient. These asymptotic expressions for μ and ν are substantially different from the expressions given by Prandtl for a 'steady' disturbance,† viz.

$$\mu = c^{-2}, \quad \nu = c, \tag{4.3.21}$$

† A steady disturbance, in terms of our analysis, is a Fourier component with a wave-number such that $\kappa_1 = 0$; putting $\kappa_1 = 0$, $e_2 = e_3 = c^{-\frac{1}{2}}$, in (4.3.11) and the two similar equations and using the continuity condition, we find

$$dZ_1''(0, c^{\frac{1}{2}}\kappa_2, c^{\frac{1}{2}}\kappa_3) = \frac{1}{c}\, dZ_1'(0, \kappa_2, \kappa_3), \quad d\begin{Bmatrix} Z_2'' \\ Z_3'' \end{Bmatrix}(0, c^{\frac{1}{2}}\kappa_2, c^{\frac{1}{2}}\kappa_3) = c^{\frac{1}{2}} d\begin{Bmatrix} Z_2' \\ Z_3' \end{Bmatrix}(0, \kappa_2, \kappa_3),$$

in agreement with (4.3.21).

which have often been used to predict the turbulence changes in wind tunnels in lieu of a more accurate theory.

As mentioned earlier in the section, not many of the contractions used in wind tunnels will be sufficiently rapid to satisfy the criterion (4.3.2) for the linearized theory to be applicable. No accounts of measurements of turbulent-velocity fluctuations made on both the upstream and downstream sides of a sudden distortion appear to have been published, so that it is not yet possible to test the soundness of the theory. The theory is of interest for its own sake as a demonstration of how completely—and easily—a problem of turbulent motion may be solved when the relevant equations are linear.

CHAPTER V

THE GENERAL DYNAMICS OF DECAY

Chapters II and III were concerned with the mathematical equipment needed to represent the turbulent motion at one instant in a statistical manner, taking into account the continuity condition and any symmetry conditions which may exist. The essence of our problem is to determine the variation of this statistical representation with the passage of time. In this chapter we shall consider the general method of going about the problem and shall investigate some properties of the decay that follow immediately from the dynamical equations. These exact deductions, which are severely limited in scope, will be supplemented in later chapters with intuitive hypotheses about the nature of the decay.

5.1. Methods of using the Navier-Stokes equation

The equation governing the variation of the spatial distribution of the velocity with time is the Navier-Stokes equation

$$\frac{\partial \mathbf{u}(\mathbf{x},t)}{\partial t} = -\mathbf{u}.\nabla\mathbf{u} - \frac{1}{\rho}\nabla p + \nu\nabla^2\mathbf{u}. \qquad (5.1.1)$$

There appear to be two distinct courses open to us in making use of this equation in an investigation of turbulent motion, the first being as follows.

Equation (5.1.1) can be regarded as determining the distribution of velocity over all space at time t in terms of the distribution over space at some initial instant t_0. In other words (5.1.1) determines the development, with time, of each realization of the turbulent field. Provided that the solution of (5.1.1) can be obtained explicitly, the value of any statistical quantity at time t can then be obtained by averaging (the appropriate function of) this explicit solution for $\mathbf{u}$ over the whole ensemble of realized fields, that is, over the ensemble of initial spatial distributions of $\mathbf{u}$.

The success of this direct approach evidently depends on the possibility of being able to solve the equation (5.1.1) explicitly. If the initial conditions are given numerically (as against being given in probability) it is of course possible to solve (5.1.1) for a realized field by a numerical step-by-step integration with respect to t. Such a procedure is very laborious and, from a theoretical point of view, seems unlikely to reveal the fundamental features of a statistical problem. Nevertheless, the lack (as yet) of a wholly successful alternative method may make it worth while to use direct calculation of realized fields in one or two sample cases to provide a guide to the solution of the problem. There are ways in which the labour required may be reduced. For instance, if the calculation is carried out for a large region of a single realized field, it may be possible to compute averages of the required quantities over space and to use ergodic theory to relate these space averages to probability averages. A first attempt at a solution along these lines has been made by H. W. Emmons (1947) for the hypothetical case of a two-dimensional turbulent flow between two fixed parallel planes, the initial spatial distribution of velocity being chosen to have arbitrary fluctuations superposed on the expected distribution of mean velocity.

A solution of (5.1.1) in closed form has not yet been found, but it is possible by an iteration process to compute $\mathbf{u}(\mathbf{x}, t)$ as a Taylor's series in $t-t_0$ (assuming it to exist), viz.

$$\mathbf{u}(\mathbf{x}, t) = \mathbf{u}(\mathbf{x}, t_0) + (t-t_0)\left[\frac{\partial \mathbf{u}(\mathbf{x}, t)}{\partial t}\right]_{t_0} + \frac{1}{2!}(t-t_0)^2\left[\frac{\partial^2 \mathbf{u}(\mathbf{x}, t)}{\partial t^2}\right]_{t_0} + \dots, \tag{5.1.2}$$

in which the coefficients are functions of $\mathbf{u}(\mathbf{x}, t_0)$ (and its derivatives with respect to $\mathbf{x}$) only. The coefficient of the term of first degree is given by (5.1.1) directly. The pressure p occurs in (5.1.1) but may be eliminated with the aid of the equation obtained by taking the divergence of (5.1.1), viz.

$$\frac{1}{\rho}\nabla^2 p = -\frac{\partial^2 u_i u_j}{\partial x_i \partial x_j}; \tag{5.1.3}$$

in an infinite fluid only the particular solution of (5.1.3) is relevant, so that

$$\frac{1}{\rho}p(\mathbf{x}) = \frac{1}{4\pi}\int \frac{\partial^2 u_i'' u_j''}{\partial x_i'' \partial x_j''}\frac{d\mathbf{x}''}{|\mathbf{x}''-\mathbf{x}|}. \tag{5.1.4}$$

The next coefficient is

$$\left(\frac{\partial^2 \mathbf{u}}{\partial t^2}\right)_{t_0} = \left(-\frac{\partial \mathbf{u}}{\partial t}.\nabla\mathbf{u} - \mathbf{u}.\nabla\frac{\partial \mathbf{u}}{\partial t} - \frac{1}{\rho}\nabla\frac{\partial p}{\partial t} + \nu\nabla^2\frac{\partial \mathbf{u}}{\partial t}\right)_{t_0}. \tag{5.1.5}$$

By eliminating p and $\partial \mathbf{u}/\partial t$ with the aid of (5.1.4) and (5.1.1) respectively, the right side of (5.1.5) can thus be expressed in terms of $\mathbf{u}(\mathbf{x}, t_0)$ and its derivatives with respect to $\mathbf{x}$, and a similar procedure is capable of determining the general derivative

$$\left(\frac{\partial^n \mathbf{u}}{\partial t^n}\right)_{t_0}.$$

It will be noted that the coefficient of $(t-t_0)^n$ in the power series contains terms of the rth degree of $\mathbf{u}(\mathbf{x}, t_0)$, where r takes all values from 1 to $n+1$.

Consequently if the initial conditions are given in probability—that is, if the joint-probability distribution of $\mathbf{u}(\mathbf{x}, t_0)$ at any n values of $\mathbf{x}$ is given, for any value of n—the power-series solution (5.1.2) can be used to determine the mean value of any required function of the velocity at time t. For instance, an m-order velocity-product mean value at time t is determined as a power series in $t-t_0$, the coefficient of the term of zero degree being an m-order product mean value at time t_0 and succeeding coefficients being product mean values of m and higher orders. In general this direct method is not of much practical value, on account of the great complexity of all but the first two or three terms of the series. Its value lies perhaps in the direct demonstration that the accuracy with which the value of an m-order product mean value at time t can be calculated depends on the number of product mean values of different order which are known at time t_0; for values of $t-t_0$ small enough for a linear approximation to be adequate the values of m- and $(m+1)$-order mean values at time t_0 are sufficient to determine accurately the m-order mean value at time t, while for slightly larger values of $t-t_0$ a knowledge of $(m+2)$-order mean values at time t_0 is needed also, and so on. The greater the value of $t-t_0$, the more of the complete probability distribution of $\mathbf{u}(\mathbf{x}, t_0)$ must we know in order to calculate any mean value at time t.†

The second method of making use of equation (5.1.1) is to convert it into a set of equations for the variation of statistical quantities with time. This is done by making some suitable operation

† This conclusion points firmly to the need for a method of specifying the complete probability distribution of $\mathbf{u}(\mathbf{x}, t_0)$ (as a random function of $\mathbf{x}$) in a form which would allow an attempt to determine, from (5.1.1), its time dependence as a whole, and not as a collection of mean values.

on the equation in order to get a time differential of the required quantity on the left side and then taking a probability average of both sides. The resulting equations, one for each of the infinite number of quantities required to specify the velocity field completely in the statistical sense, are then to be solved in terms of the statistical specification of the turbulence at the initial instant t_0. As with the other method of using the dynamical equation, we are particularly interested in obtaining a solution for general initial conditions in order that an approach to a statistical state partially independent of the initial conditions may (if it exists) be investigated.

This plan of averaging the differential equations is the one that has generally been adopted in past research. We can illustrate the procedure by formulating the equation for time variation of the velocity correlation $R_{ij}(\mathbf{r}, t)$. If $\mathbf{u}'$ and p' denote the velocity and pressure at the point $\mathbf{x}' = \mathbf{x} + \mathbf{r}$, the Navier-Stokes equations at the points $\mathbf{x}$ and $\mathbf{x}'$ can be written as

$$\frac{\partial u_i}{\partial t} = -\frac{\partial u_i u_k}{\partial x_k} - \frac{1}{\rho}\frac{\partial p}{\partial x_i} + \nu \nabla_{\mathbf{x}}^2 u_i, \tag{5.1.6}$$

$$\frac{\partial u_j'}{\partial t} = -\frac{\partial u_j' u_k'}{\partial x_k'} - \frac{1}{\rho}\frac{\partial p'}{\partial x_j'} + \nu \nabla_{\mathbf{x}'}^2 u_j'. \tag{5.1.7}$$

On multiplying the first of these equations by u_j' and the second by u_i and adding the two, we obtain the time rate of change of $u_i u_j'$, of which the probability average is

$$\frac{\partial \overline{u_i u_j'}}{\partial t} = -\left(\overline{u_i \frac{\partial u_j' u_k'}{\partial x_k'}} + \overline{u_j' \frac{\partial u_i u_k}{\partial x_k}}\right) - \frac{1}{\rho}\left(\overline{u_i \frac{\partial p'}{\partial x_j'}} + \overline{u_j' \frac{\partial p}{\partial x_i}}\right)$$
$$+ \nu\left(\overline{u_i \nabla_{\mathbf{x}'}^2 u_j'} + \overline{u_j' \nabla_{\mathbf{x}}^2 u_i}\right). \tag{5.1.8}$$

The operations of differentiation and taking an average permute; u_i is independent of $\mathbf{x}'$, so that $\overline{u_i \frac{\partial u_j' u_k'}{\partial x_k'}} = \frac{\partial \overline{u_i u_j' u_k'}}{\partial x_k'}$, and similarly for other terms; and lastly, $\partial/\partial x_i$ and $\partial/\partial x_i'$ can be replaced by $-\partial/\partial r_i$ and $\partial/\partial r_i$ respectively when they operate on a two-point mean value. With all these aids, (5.1.8) becomes

$$\frac{\partial R_{ij}(\mathbf{r}, t)}{\partial t} = \frac{\partial}{\partial r_k}\left(\overline{u_i u_k u_j'} - \overline{u_i u_k' u_j'}\right) + \frac{1}{\rho}\left(\frac{\partial \overline{p u_j'}}{\partial r_i} - \frac{\partial \overline{p' u_i}}{\partial r_j}\right) + 2\nu \nabla^2 R_{ij}(\mathbf{r}, t), \tag{5.1.9}$$

where ∇^2 now has the meaning of $\partial^2/(\partial r_l \partial r_l)$. The right side of (5.1.9) can be written entirely in terms of velocity-product mean values, since we have, from (5.1.4),

$$\frac{1}{\rho}\overline{pu_j'} = \frac{1}{4\pi}\int \frac{1}{|\mathbf{x}''-\mathbf{x}|}\frac{\partial^2 \overline{u_i''u_k''u_j'}}{\partial x_i''\partial x_k''}d\mathbf{x}''$$

$$= \frac{1}{4\pi}\int \frac{1}{|\mathbf{r}-\mathbf{s}|}\frac{\partial \overline{u_i''u_k''u_j'}}{\partial s_i \partial s_k}d\mathbf{s}, \qquad (5.1.10)$$

where u_j'' is the velocity at the point $\mathbf{x}''=\mathbf{x}'-\mathbf{s}$, and the integration is over all $\mathbf{s}$-space; a similar expression can be found for $\frac{1}{\rho}\overline{p'u_i}$.

The plan is now to solve this dynamical equation for $R_{ij}(\mathbf{r}, t)$, for given values of R_{ij} as a function of $\mathbf{r}$ (preferably expressed in general form) at some initial instant t_0. A difficulty in principle is immediately evident from the form of (5.1.9). The right side contains velocity-product mean values other than $R_{ij}(\mathbf{r}, t)$, so that by itself the equation is insufficient to determine $R_{ij}(\mathbf{r}, t)$. This situation has of course been brought about by the non-linearity of the Navier-Stokes equation. The nature of the difficulties created by the non-linearity has been changed by the operation of taking an average of the equation; whereas we were originally faced with a non-linear integro-differential equation (5.1.1) in a single dependent variable $\mathbf{u}(\mathbf{x}, t)$ (after p had been eliminated with the aid of the continuity equation), we now have a set of simultaneous linear integro-differential equations (of which that for $R_{ij}(\mathbf{r}, t)$, viz. (5.1.9), is typical) for the velocity-product mean values.

The essence of the difficulty with this second method of attack is that we do not know how to get a sufficient set of simultaneous equations for the velocity-product mean values. If we form the dynamical equation for the third-order three-point product mean value $\overline{u_i u_j' u_k''}$, a special case of which occurs on the right side of (5.1.9), we find that it contains fourth-order product mean values. A continuation of this process leads merely to an infinite set of equations, the number of equations being always one fewer than the number of dependent variables. The explanation of the position appears to be that the product mean values contribute less and less information to the complete probability distribution as their order increases, so that the set of equations becomes complete asymptotically. Product mean values of high order correspond to integral moments of the probability distribution function of high order, and although an infinite number of such moments is needed to determine the distribution *exactly*, we obtain an increasingly good approximation to it by specifying a larger and larger number of the integral moments. We could represent the turbulence statistically to a good approximation by a large but finite number N of velocity-product mean values, and a set of dynamical equations (like that described above), one for each of the N product mean values, would then be sufficient—with certain restrictions on the accuracy—to determine the development with time, since we should be at liberty to choose an arbitrary value (at all values of t) for the $(N+1)$-order product mean value which occurs in the last equation of the set.† This arbitrary choice of the $(N+1)$-order product mean will have an immediate effect on the rate of change of the N-order mean, and will affect lower order means at successively later times; consequently accurate calculation of the development of the turbulence with time is possible for a time interval which is greater, the greater is the number of product mean values which are given at the initial instant.

It is now clear that there is no essential difference between, on the one hand, the above procedure of averaging the dynamical equation for the various velocity products, and, on the other hand, the procedure described earlier of obtaining the solution of the Navier-Stokes equation as a power series in $t-t_0$ and averaging the appropriate function of this series solution. In both cases the time-development of a partial statistical representation of the turbulence can be calculated over a time interval which increases with the completeness of the statistical representation. Two mathematical contributions (possibly unattainable) would greatly increase our ability to analyse the decay of the turbulence. One is the deduction of a general solution of the Navier-Stokes equation

† An equivalent and perhaps better plan would be to assume that for all values of t the $(N+1)$-order product mean is the same function of lower order product means as for an arbitrarily chosen probability distribution. M. Millionshtchikov (1941*a*), in a special context, made the convenient assumption (which may also be accurate, see § 8.2) that fourth-order product means are related to lower-order product means in the same way as for a *normal* probability distribution, and solved the resulting set of two dynamical equations.

(5.1.1), in *closed* form, for an arbitrary initial spatial distribution of the velocity, which would render the power series solution unnecessary. The other is the formulation of a *single* dynamical equation describing the time-variation of the complete probability distribution of the velocity field, which would render unnecessary the solving of a large number of dynamical equations for product mean values. The latter task encounters the fundamental problem of the representation of a statistical distribution in function space.†

5.2. The flow of energy

It was seen, in Chapter II, that a description of the velocity distribution by means of a Fourier analysis makes possible the concept of mechanical components, or degrees of freedom, which make additive contributions to the energy of the motion. We shall obtain a partial understanding of the general process of decay if we can see, qualitatively, how the energy associated with the various degrees of freedom varies with time. In general, there is a flow of energy from one degree of freedom to another, from one (directional) component of the velocity to another, and a flow out of the mechanical system into heat. All these changes are comprehended in a description of the temporal variation of the Fourier coefficient $\mathrm{d}\mathbf{Z}(\boldsymbol{\kappa})$ or of the spectrum tensor $\Phi_{ij}(\boldsymbol{\kappa})$. The examination of both of these quantities proves to be useful.

The equation for the rate of change of $\mathrm{d}\mathbf{Z}(\boldsymbol{\kappa})$ is obtained from the Navier-Stokes equation (5.1.1) by multiplying by

$$\frac{1}{(2\pi)^3}\,e^{-\iota\boldsymbol{\kappa}.\mathbf{x}}\left(\frac{e^{-\iota \mathrm{d}\kappa_1 x_1}-1}{-\iota x_1}\right)\left(\frac{e^{-\iota \mathrm{d}\kappa_2 x_2}-1}{-\iota x_2}\right)\left(\frac{e^{-\iota \mathrm{d}\kappa_3 x_3}-1}{-\iota x_3}\right)$$

and integrating over all $\mathbf{x}$-space in the manner of (2.5.2). We find, neglecting terms of higher order in $\mathrm{d}\boldsymbol{\kappa}$,

$$\frac{\partial \mathrm{d}\mathbf{Z}(\boldsymbol{\kappa})}{\partial t} = -\iota\int_{\boldsymbol{\kappa}'}\boldsymbol{\kappa}'.\mathrm{d}\mathbf{Z}(\boldsymbol{\kappa}-\boldsymbol{\kappa}')\,\mathrm{d}\mathbf{Z}(\boldsymbol{\kappa}') - \iota\boldsymbol{\kappa}\,\mathrm{d}W(\boldsymbol{\kappa}) - \nu\kappa^2\mathrm{d}\mathbf{Z}(\boldsymbol{\kappa}), \tag{5.2.1}$$ ‡

† Note added in proof: For a recent paper which is important in this connexion, see E. Hopf, 'Statistical hydromechanics and functional calculus', *J. Rat. Mech. Anal.* 1, 1952, 87.

‡ We shall not show explicitly the dependence of quantities like $\mathrm{d}\mathbf{Z}(\boldsymbol{\kappa})$ on t in this chapter, so long as there is no risk of confusion, in order to avoid too great an array of symbols in the equations.

where $dW(\boldsymbol{\kappa})$ is the Fourier coefficient of the pressure $p(x)/\rho$ in the sense that $d\mathbf{Z}(\boldsymbol{\kappa})$ is that of $\mathbf{u}(\mathbf{x})$, and Parseval's formula (see footnote to p. 56) has been used (formally) to obtain the transform of the non-linear term. In view of the relation (5.1.4) for the pressure, we have, using Parseval's formula again,

$$dW(\boldsymbol{\kappa}) = -\frac{1}{\kappa^2}\int_{\boldsymbol{\kappa}'} \boldsymbol{\kappa}.d\mathbf{Z}(\boldsymbol{\kappa}-\boldsymbol{\kappa}')\,\boldsymbol{\kappa}.d\mathbf{Z}(\boldsymbol{\kappa}'). \qquad (5.2.2)$$

Hence, with the use of the continuity relation in the form

$$\boldsymbol{\kappa}.d\mathbf{Z}(\boldsymbol{\kappa}-\boldsymbol{\kappa}') = \boldsymbol{\kappa}'.d\mathbf{Z}(\boldsymbol{\kappa}-\boldsymbol{\kappa}'), \qquad (5.2.3)$$

(5.2.1) becomes

$$\frac{\partial d\mathbf{Z}(\boldsymbol{\kappa})}{\partial t} = \iota\int_{\boldsymbol{\kappa}'} \boldsymbol{\kappa}.d\mathbf{Z}(\boldsymbol{\kappa}-\boldsymbol{\kappa}')\left[-d\mathbf{Z}(\boldsymbol{\kappa}') + \frac{\boldsymbol{\kappa}}{\kappa^2}\boldsymbol{\kappa}.d\mathbf{Z}(\boldsymbol{\kappa}')\right] - \nu\kappa^2 d\mathbf{Z}(\boldsymbol{\kappa}). \qquad (5.2.4)$$

This dynamical equation for the Fourier coefficient $d\mathbf{Z}(\boldsymbol{\kappa})$ tells us neither more nor less than the Navier-Stokes equation from which it is obtained, but it can perhaps be more readily interpreted in terms of analytical dynamics.

The rate of change of the energy associated with the Fourier coefficient $d\mathbf{Z}(\boldsymbol{\kappa})$ is found from (5.2.4) to be described by

$$\begin{aligned}\frac{\partial\, dZ_i^*(\boldsymbol{\kappa})\,dZ_j(\boldsymbol{\kappa})}{\partial t} &= \iota\int_{\boldsymbol{\kappa}'}[\boldsymbol{\kappa}.d\mathbf{Z}^*(\boldsymbol{\kappa}-\boldsymbol{\kappa}')\,dZ_i^*(\boldsymbol{\kappa}')\,dZ_j(\boldsymbol{\kappa}) - \boldsymbol{\kappa}.d\mathbf{Z}(\boldsymbol{\kappa}-\boldsymbol{\kappa}')\,dZ_i^*(\boldsymbol{\kappa})\,dZ_j(\boldsymbol{\kappa}')]\\ &\quad + \iota\int_{\boldsymbol{\kappa}'}\frac{1}{\kappa^2}[\boldsymbol{\kappa}.d\mathbf{Z}(\boldsymbol{\kappa}-\boldsymbol{\kappa}')\,\boldsymbol{\kappa}.d\mathbf{Z}(\boldsymbol{\kappa}')\,\kappa_j\,dZ_i^*(\boldsymbol{\kappa})\\ &\quad - \boldsymbol{\kappa}.d\mathbf{Z}^*(\boldsymbol{\kappa}-\boldsymbol{\kappa}')\,\boldsymbol{\kappa}.d\mathbf{Z}^*(\boldsymbol{\kappa}')\,\kappa_i\,dZ_j(\boldsymbol{\kappa})] - 2\nu\kappa^2 dZ_i^*(\boldsymbol{\kappa})\,dZ_j(\boldsymbol{\kappa}). \qquad (5.2.5)\end{aligned}$$

The three terms on the right side represent the effects of inertia, pressure,† and viscous forces respectively. The linear viscous forces present no difficulty, and if no other forces were present $d\mathbf{Z}(\boldsymbol{\kappa})$ would decrease with time as $e^{-\nu\kappa^2 t}$; viscous forces change the amplitudes, but not the phases,‡ of the Fourier coefficients. The

† 'Pressure' forces are a particular manifestation of inertia forces in a fluid which fills all space, as is clear from the relation (5.1.4). However, it is convenient on occasions to treat pressure and inertia forces as distinct inasmuch as they produce different effects.

‡ Each of the directional components of $d\mathbf{Z}(\boldsymbol{\kappa})$ is complex and can be expressed in the form $|dZ_i(\boldsymbol{\kappa})|\,e^{\iota\theta_i(\boldsymbol{\kappa})}$ (no summation over the values of i), $\theta_i(\boldsymbol{\kappa})$ being the 'phase' of the i-component.

effects of the inertia and pressure forces, on the other hand, are represented by non-linear terms of considerable complexity. The appearance of the integrals in (5.2.4) and (5.2.5) demonstrates the continual modulation, or interaction, which occurs between every pair of wave-numbers during the decay. The (multiplicative) interaction of two Fourier components with wave-numbers $\boldsymbol{\kappa}'$ and $\boldsymbol{\kappa}''$ forms a Fourier component with wave-number $\boldsymbol{\kappa}'+\boldsymbol{\kappa}''$, so that points in wave-number space can be regarded as being linked with each other, through the inertia forces, in groups of three.

One or two features of the coupling between the various degrees of freedom of the system can be deduced from (5.2.5). First, one of the properties of the function $d\mathbf{Z}(\boldsymbol{\kappa})$, as defined by (2.5.1) and (2.5.2), is that $d\mathbf{Z}^*(\boldsymbol{\kappa}) = d\mathbf{Z}(-\boldsymbol{\kappa})$, from which it follows that the contribution to $\frac{\partial}{\partial t}\int_{\boldsymbol{\kappa}} dZ_i^*(\boldsymbol{\kappa})\, dZ_j(\boldsymbol{\kappa})$ from the effect of inertia forces is zero. The effect of inertia forces is thus to transfer energy from one part of wave-number space to another, without changing the total amount of energy associated with any directional component of the energy. Secondly, we see, with the use of the continuity relation (2.5.7), that the contribution to $\frac{\partial}{\partial t} dZ_i^*(\boldsymbol{\kappa})\, dZ_i(\boldsymbol{\kappa})$ from the effect of pressure forces is zero. The effect of pressure forces is evidently to transfer energy (at a rate depending on the interaction with all other wave-numbers) from one directional component of $d\mathbf{Z}(\boldsymbol{\kappa})$ to another. These consequences of pressure and inertia forces are net effects, in the sense that, for example, pressure forces produce some interaction between $d\mathbf{Z}(\boldsymbol{\kappa})$ and $d\mathbf{Z}(\boldsymbol{\kappa}')$ (for see (5.2.4)), although the total effect of such interactions for all values of $\boldsymbol{\kappa}'$ is such as to leave $dZ_i^*(\boldsymbol{\kappa})\, dZ_i(\boldsymbol{\kappa})$ unaltered. The exchanges of energy are dependent, in general, on the relations between the phases of the different Fourier components as well as on their amplitudes, and it is in the elucidation of the average properties of the phase relations that the key to the determination of the energy spectrum during the decay lies.

Only the average properties of the flow of energy interest us, since the flow will fluctuate with different realizations (although the fluctuations must be consistent with the general properties established in the preceding paragraph). To obtain average quantities which are simply related to velocity-product mean values, we must make an appropriate limiting operation on (5.2.5). For instance, if we contract the indices i and j in (5.2.5) in order to obtain the physically important case of total energy, divide by $d\boldsymbol{\varkappa}$, and proceed to the limit $d\boldsymbol{\varkappa} \to 0$, we obtain (in view of (2.5.5))

$$\frac{\partial \frac{1}{2}\Phi_{ii}(\boldsymbol{\varkappa})}{\partial t} = \int Q(\boldsymbol{\varkappa}, \boldsymbol{\varkappa}')\, d\boldsymbol{\varkappa}' - \nu\kappa^2 \Phi_{ii}(\boldsymbol{\varkappa}), \qquad (5.2.6)$$

where the limit

$$Q(\boldsymbol{\varkappa}, \boldsymbol{\varkappa}') = \lim_{d\boldsymbol{\varkappa},\, d\boldsymbol{\varkappa}' \to 0} \frac{\iota}{2} \left[\frac{\overline{\boldsymbol{\varkappa} . d\mathbf{Z}^*(\boldsymbol{\varkappa} - \boldsymbol{\varkappa}')\, d\mathbf{Z}^*(\boldsymbol{\varkappa}') . d\mathbf{Z}(\boldsymbol{\varkappa})}}{d\boldsymbol{\varkappa}\, d\boldsymbol{\varkappa}'} - \frac{\overline{\boldsymbol{\varkappa} . d\mathbf{Z}(\boldsymbol{\varkappa} - \boldsymbol{\varkappa}')\, d\mathbf{Z}^*(\boldsymbol{\varkappa}) . d\mathbf{Z}(\boldsymbol{\varkappa}')}}{d\boldsymbol{\varkappa}\, d\boldsymbol{\varkappa}'} \right] \qquad (5.2.7)$$

may be shown to exist.† We may interpret $Q(\boldsymbol{\varkappa}, \boldsymbol{\varkappa}')$ (with caution, since (5.2.6) is capable of defining only the net flow of energy into $d\boldsymbol{\varkappa}$ from *all* parts of wave-number space) as the net mean rate of transfer of energy from the volume element $d\boldsymbol{\varkappa}'$ in wave-number space to the element $d\boldsymbol{\varkappa}$. Owing to the occurrence of the phases of the Fourier components in the expression for $Q(\boldsymbol{\varkappa}, \boldsymbol{\varkappa}')$ (in contradistinction to $\Phi_{ii}(\boldsymbol{\varkappa})$ which depends on amplitudes alone), it is not easy to get a clear picture of the mechanical processes responsible for the transfer of energy.

We readily verify that (5.2.7) satisfies the conservation law

$$Q(\boldsymbol{\varkappa}, \boldsymbol{\varkappa}') + Q(\boldsymbol{\varkappa}', \boldsymbol{\varkappa}) = 0, \qquad (5.2.8)$$

which reflects the fact that in an infinite fluid free from boundaries inertia forces can do no more than transfer energy. The rate of

† It will be recalled that it was assumed in § 2.4 that the energy of the turbulence is spread continuously over the spectrum, which is equivalent to assuming that $\lim_{d\boldsymbol{\varkappa} \to 0} \dfrac{\overline{dZ_i^*(\boldsymbol{\varkappa})\, dZ_i(\boldsymbol{\varkappa})}}{d\boldsymbol{\varkappa}}$ exists. If this assumption were not valid at some instant, and for some value of $\boldsymbol{\varkappa}$, owing to the occurrence of a line or step in the spectrum (not to be confused with a steady periodic distribution of the *mean* velocity) with a corresponding large magnitude of $d\mathbf{Z}(\boldsymbol{\varkappa})$, the limit (5.2.7) likewise would not exist. The form of the dynamical equation (5.2.4) shows that the rate of change of $|d\mathbf{Z}|^2$ is of the *third* degree in $|d\mathbf{Z}|$, the consequence of which is that the value of $|d\mathbf{Z}|$ at any finite time later would be reduced to an order such that the limits (2.5.5) and (5.2.7) exist. Spectral lines containing a finite amount of energy can exist instantaneously, if produced by some external means, but they cannot persist when the turbulence is left to itself.

change of total energy per unit mass of fluid is thus

$$\frac{d\frac{1}{2}\overline{u_i u_i}}{dt} = \frac{d}{dt}\int_0^\infty E(\kappa)\,d\kappa$$

$$= \frac{d}{dt}\int \tfrac{1}{2}\Phi_{ii}(\boldsymbol{\kappa})\,d\boldsymbol{\kappa} = -\nu\int \kappa^2\Phi_{ii}(\boldsymbol{\kappa})\,d\boldsymbol{\kappa} \qquad \text{from (5.2.6)},$$

$$= -\nu\int \Omega_{ii}(\boldsymbol{\kappa})\,d\boldsymbol{\kappa} = -\nu\overline{\omega_i\omega_i} \qquad \text{from (3.2.4)},$$

and

$$= -2\nu\int_0^\infty \kappa^2 E(\kappa)\,d\kappa \tag{5.2.9}$$

from (3.1.2) and (3.1.5), and is determined by the average dispersion of the energy about the origin in wave-number space. (5.2.6) also shows that small-scale Fourier components lose energy by viscous dissipation more rapidly than large-scale components. However, this differential decay does not always lead to rapid changes in the shape of the energy spectrum, since the inertial transfer of energy between different wave-numbers tends to fill up any regions of low-energy density in wave-number space. If it were not for the higher rate of dissipation at larger values of κ the energy would tend to spread itself over an infinite range of wave-numbers, and there would be an 'escape' of energy to infinity.

The mean values of the non-linear terms on the right side of (5.2.5) can be expressed in terms of velocity-product mean values, as could be established directly from the relation between $d\mathbf{Z}(\boldsymbol{\kappa})$ and $\mathbf{u}(\mathbf{x})$. It is simpler to return to (5.1.9) and take its Fourier transform, giving the two complementary dynamical equations:

$$\frac{\partial R_{ij}(\mathbf{r})}{\partial t} = T_{ij}(\mathbf{r}) + P_{ij}(\mathbf{r}) + 2\nu\nabla^2 R_{ij}(\mathbf{r}), \tag{5.2.10}$$

$$\frac{\partial \Phi_{ij}(\boldsymbol{\kappa})}{\partial t} = \Gamma_{ij}(\boldsymbol{\kappa}) + \Pi_{ij}(\boldsymbol{\kappa}) - 2\nu\kappa^2\Phi_{ij}(\boldsymbol{\kappa}), \tag{5.2.11}$$

where

$$T_{ij}(\mathbf{r}) = \frac{\partial}{\partial r_k}(\overline{u_i u_k u_j'} - \overline{u_i u_k' u_j'}) = \int \Gamma_{ij}(\boldsymbol{\kappa})\,e^{i\boldsymbol{\kappa}.\mathbf{r}}\,d\boldsymbol{\kappa}, \tag{5.2.12}$$

$$P_{ij}(\mathbf{r}) = \frac{1}{\rho}\left(\frac{\partial\overline{pu_j'}}{\partial r_i} - \frac{\partial\overline{p'u_i}}{\partial r_j}\right) = \int \Pi_{ij}(\boldsymbol{\kappa})\,e^{i\boldsymbol{\kappa}.\mathbf{r}}\,d\boldsymbol{\kappa}. \tag{5.2.13}$$

The three terms on the right side of (5.2.11) represent the contributions of inertia, pressure and viscous forces respectively to the rate of change of $\Phi_{ij}(\boldsymbol{\kappa})$, and are identical with the three terms on the right side of (5.2.5) after the latter have been divided by $d\boldsymbol{\kappa}$ and the limit as $d\boldsymbol{\kappa} \to 0$ has been taken. The properties already established for the terms in (5.2.5) can also be established for those of (5.2.11), although the proofs are superficially unrelated. Taking first the contribution from inertia forces, we note that

$$T_{ij}(0) = \overline{u_i u_k \frac{\partial u_j}{\partial x_k}} + \overline{\frac{\partial u_i}{\partial x_k} u_k u_j} = \frac{\partial}{\partial x_k} \overline{u_i u_j u_k} = 0,$$

that is,

$$\int \Gamma_{ij}(\boldsymbol{\kappa})\, d\boldsymbol{\kappa} = 0, \tag{5.2.14}$$

showing, as expected, that the rate of change of

$$\int \Phi_{ij}(\boldsymbol{\kappa})\, d\boldsymbol{\kappa} = \overline{u_i u_j},$$

owing to the effect of inertia forces is zero; inertia forces alter the distribution of density, in wave-number space, of contributions to $\overline{u_i u_j}$, but they leave unchanged the total contribution. For the pressure forces we have

$$P_{ii}(\mathbf{r}) = 0$$

for all values of $\mathbf{r}$, from the condition of incompressibility, so that

$$\Pi_{ii}(\boldsymbol{\kappa}) = 0$$

for all values of $\boldsymbol{\kappa}$; the net effect of pressure forces is thus to conserve the total energy contributed by any small region of wave-number space, although the directional distribution of this energy might be changed.

Although it has been shown that inertia forces produce a flow of energy between different wave-numbers for the same velocity component, and that pressure forces transfer energy between different velocity components for the same wave-number, nothing has been established about the directions of these energy transfers. Definite directions of the energy transfers, valid under all conditions, are not to be expected, since special initial values of the phases of the Fourier components could always be devised to produce, temporarily, any desired flow of energy. Hence the most that we can do is to deduce general trends in the energy flow with the aid of rough physical arguments. The effect of inertia forces is to spread energy over an increasingly wide range of wave-numbers, and thus to direct (in a statistical sense) energy towards those parts of wave-number space where the energy density is lowest. Since viscous dissipation is more rapid at larger wave-numbers, the trend of the inertial transfer of energy will almost always be from small to large wave-numbers, that is, small eddies† will derive energy from large eddies. A similarly simple argument suggests that the transfer of energy by pressure forces tends to equalize the mean squares of the directional components of the velocity (and consequently, in view of the continuity condition, to distribute the energy associated with the volume element $d\boldsymbol{\kappa}$ of wave-number space uniformly over directions normal to $\boldsymbol{\kappa}$). A positive fluctuation in the pressure at any point in the fluid represents, in a sense, stored kinetic energy, and the velocity component which has the largest mean square will make the largest average contribution to the pressure fluctuations. But in releasing the stored kinetic energy the pressure is non-directional, and the probable consequence is that it builds up the weaker velocity component at the expense of the stronger. The argument can be given more analytical form when the turbulence has axisymmetry (Batchelor, 1946), but remains essentially as stated. The conclusion that in general homogeneous turbulence tends to a state of uniform directional distribution of energy (and, we may speculate, to a state of isotropy) is in agreement with experiment and has long been known empirically.

5.3. The permanence of big eddies

It is possible to examine the flow of energy in more detail in the neighbourhood of $\kappa = 0$, by expanding the various terms of (5.2.11) as power series‡ in the components of $\boldsymbol{\kappa}$, as has already been done for $\Phi_{ij}(\boldsymbol{\kappa})$ in § 3.1. The result established there (see (3.1.9)) is that

$$\Phi_{ij}(\boldsymbol{\kappa}) = \kappa_l \kappa_m C_{ijlm} + O(\kappa^3), \tag{5.3.1}$$

† 'Small eddy' is used here and elsewhere as a concise term for a Fourier component belonging to a small length scale, or large wave-number.

‡ Again making the assumption that the first few derivatives of the required functions at $\kappa = 0$ exist. It seems safe to make such an assumption about any statistical quantity, but not about unaveraged quantities involving the Fourier coefficient $d\mathbf{Z}(\boldsymbol{\kappa})$ which is not an analytic function of $\boldsymbol{\kappa}$. But see the note on p. 195.

where C_{ijlm} is a tensor which is independent of $\boldsymbol{\kappa}$ but may depend on t. We wish to examine the corresponding terms of lowest order in the expansions of $\Gamma_{ij}(\boldsymbol{\kappa})$ and $\Pi_{ij}(\boldsymbol{\kappa})$ (the viscosity term requiring no further attention).

Taking $\Gamma_{ij}(\boldsymbol{\kappa})$ first, we define $\Upsilon_{ikj}(\boldsymbol{\kappa})$ as the Fourier transform of $\overline{u_i u_k u_j'}$, so that

$$\overline{u_i u_k u_j'} = \int \Upsilon_{ikj}(\boldsymbol{\kappa})\, e^{\iota\boldsymbol{\kappa}.\mathbf{r}} d\boldsymbol{\kappa} \tag{5.3.2}$$

and $$\overline{u_i u_k' u_j'} = \int \Upsilon_{jki}(\boldsymbol{\kappa})\, e^{-\iota\boldsymbol{\kappa}.\mathbf{r}}\, d\boldsymbol{\kappa} = \int \Upsilon_{jki}(-\boldsymbol{\kappa})\, e^{\iota\boldsymbol{\kappa}.\mathbf{r}}\, d\boldsymbol{\kappa}.$$

Substituting in (5.2.12) we find

$$\Gamma_{ij}(\boldsymbol{\kappa}) = \iota\kappa_k[\Upsilon_{ikj}(\boldsymbol{\kappa}) - \Upsilon_{jki}(-\boldsymbol{\kappa})]. \tag{5.3.3}$$

Now expand $\Upsilon_{ikj}(\boldsymbol{\kappa})$ as a power series in the components of $\boldsymbol{\kappa}$:

$$\Upsilon_{ikj}(\boldsymbol{\kappa}) = \Upsilon_{ikj} + \kappa_l \Upsilon_{ikjl} + \ldots,$$

where Υ_{ikj}, Υ_{ikjl}, ... are tensor functions of t alone. The continuity condition requires

$$\frac{\partial}{\partial r_j}\overline{u_i u_k u_j'} = 0,$$

i.e. $$\kappa_j \Upsilon_{ikj}(\boldsymbol{\kappa}) = 0,$$

for all values of $\boldsymbol{\kappa}$, so that the coefficients in the power series must satisfy the equations

$$\Upsilon_{ikj} = 0,$$

$$\Upsilon_{ikjl} + \Upsilon_{iklj} = 0, \tag{5.3.4}$$

etc. Hence the behaviour of $\Gamma_{ij}(\boldsymbol{\kappa})$ near $\kappa = 0$ is

$$\Gamma_{ij}(\boldsymbol{\kappa}) = \kappa_l \kappa_m \Gamma_{ijlm} + O(\kappa^3), \tag{5.3.5}$$

where $$\Gamma_{ijlm} = \tfrac{1}{2}\iota \sum_{\text{perm } i,j} \sum_{\text{perm } l,m} \Upsilon_{iljm}, \tag{5.3.6}$$

the summations being over all permutations of the indices shown.

Now let us consider $\Pi_{ij}(\boldsymbol{\kappa})$. We define $\Theta_i(\boldsymbol{\kappa})$ by

$$\frac{1}{\rho}\overline{p u_i'} = \int \Theta_i(\boldsymbol{\kappa})\, e^{\iota\boldsymbol{\kappa}.\mathbf{r}}\, d\boldsymbol{\kappa}, \tag{5.3.7}$$

from which we also have

$$\frac{1}{\rho}\overline{p' u_i} = \int \Theta_i(\boldsymbol{\kappa})\, e^{-\iota\boldsymbol{\kappa}.\mathbf{r}}\, d\boldsymbol{\kappa} = \int \Theta_i(-\boldsymbol{\kappa})\, e^{\iota\boldsymbol{\kappa}.\mathbf{r}}\, d\boldsymbol{\kappa}.$$

Substitution in (5.2.13) gives

$$\Pi_{ij}(\boldsymbol{\kappa}) = \iota[\kappa_i \Theta_j(\boldsymbol{\kappa}) - \kappa_j \Theta_i(-\boldsymbol{\kappa})]. \tag{5.3.8}$$

The expansion of $\Theta_i(\boldsymbol{\kappa})$ in powers of the components of $\boldsymbol{\kappa}$ is

$$\Theta_i(\boldsymbol{\kappa}) = \Theta_i + \kappa_l \Theta_{il} + \dots,$$

where Θ_i, Θ_{il}, ... are tensor functions of t which are required by the continuity condition, viz.

$$\kappa_i \Theta_i(\boldsymbol{\kappa}) = 0,$$

for all $\boldsymbol{\kappa}$, to satisfy the equations

$$\Theta_i = 0,$$

$$\Theta_{il} + \Theta_{li} = 0, \tag{5.3.9}$$

etc. Hence the behaviour of $\Pi_{ij}(\boldsymbol{\kappa})$ near $\kappa = 0$ is

$$\Pi_{ij}(\boldsymbol{\kappa}) = \kappa_l \kappa_m \Pi_{ijlm} + O(\kappa^3), \tag{5.3.10}$$

where
$$\Pi_{ijlm} = \tfrac{1}{2}\iota \sum_{\text{perm}\, i,j} \; \sum_{\text{perm}\, l,m} \delta_{il} \Theta_{jm}. \tag{5.3.11}$$

It seems that $\Upsilon_{ij}(\boldsymbol{\kappa})$ and $\Pi_{ij}(\boldsymbol{\kappa})$, like $\Phi_{ij}(\boldsymbol{\kappa})$, are of the second degree in κ near $\kappa = 0$. However, it is the sum of these two quantities that interests us, and one further condition concerning the sum has yet to be used. The pressure and inertia forces are related by equation (5.1.3), from which we find

$$\frac{1}{\rho} \nabla^2 \overline{p u_j'} = -\frac{\partial^2 \overline{u_i u_k u_j'}}{\partial r_i \partial r_k}.$$

An equivalent form of this equation follows from (5.3.2) and (5.3.7) as

$$\kappa^2 \Theta_j(\boldsymbol{\kappa}) = -\kappa_i \kappa_k \Upsilon_{ikj}(\boldsymbol{\kappa}). \tag{5.3.12†}$$

The leading terms of the expansions of $\Theta_j(\boldsymbol{\kappa})$ and $\Upsilon_{ikj}(\boldsymbol{\kappa})$ must therefore satisfy

$$\kappa_i \kappa_k \kappa_l (\delta_{ik} \Theta_{jl} + \Upsilon_{ikjl}) = 0$$

† It follows readily from this relation that

$$\kappa^2 \Pi_{ij}(\boldsymbol{\kappa}) = -\kappa_l [\kappa_i \Gamma_{lj}(\boldsymbol{\kappa}) + \kappa_j \Gamma_{il}(\boldsymbol{\kappa})],$$

which is a simple relation between the contributions to $\dfrac{\partial \Phi_{ij}(\boldsymbol{\kappa})}{\partial t}$ from pressure and inertia forces.

for all values of $\boldsymbol{\kappa}$, which requires

$$\sum_{\text{perm } i,k,l} (\delta_{ik}\Theta_{jl} + \Upsilon_{ikjl}) = 0. \qquad (5.3.13)$$

Now we have, from (5.3.6) and (5.3.11),

$$\Gamma_{ijlm} + \Pi_{ijlm} = \tfrac{1}{2}\iota \sum_{\text{perm } i,j} \sum_{\text{perm } l,m} (\Upsilon_{iljm} + \delta_{il}\Theta_{jm}),$$

$$= \tfrac{1}{2}\iota \sum_{\text{perm } i,j} (-\Upsilon_{lmji} - \delta_{lm}\Theta_{ji}),$$

in view of (5.3.13) and the symmetry of the tensors Υ_{iljm} and δ_{il} in the indices i and l; the relations (5.3.4) and (5.3.9) then show that

$$\Gamma_{ijlm} + \Pi_{ijlm} = 0. \qquad (5.3.14)$$

Consequently the expansion of $\Gamma_{ij}(\boldsymbol{\kappa}) + \Pi_{ij}(\boldsymbol{\kappa})$ begins with a term of not less than the *third* degree in κ.

We have seen that the whole of the right side of (5.2.11) is of order κ^3 at most when κ is small, and in view of (5.3.1) the consequence for the left side is that

$$\frac{\mathrm{d}}{\mathrm{d}t} C_{ijlm} = 0. \qquad (5.3.15)$$

In other words, in the immediate neighbourhood of $\kappa = 0$ the spectrum tensor has the same form throughout the whole history of the decay. As $\kappa \to 0$ the flow of energy owing to all three causes—inertia, pressure and viscous forces—falls off more rapidly than does the spectrum tensor itself, so that the spectrum tensor near $\kappa = 0$ has a permanent form, viz. that given it by the initial conditions. We can think of this physically as meaning that the big slow eddies interact very weakly with the remainder of the turbulence and preserve their energy intact. The directional distribution of energy in the big eddies is also permanent, so unless the turbulence is isotropic initially, the big eddies will remain anisotropic. We can foresee that it will hence be difficult in practice to generate turbulence which is completely isotropic. In general, the permanence of the distribution of energy in the big eddies is not of much practical importance because the region of the spectrum described accurately by the first term of (5.3.1) usually contains a negligible amount of energy. Nevertheless, there are some aspects of the decay which are very directly governed by the low wave-number part of the spectrum, the most striking of which is described in the next section.

The result (5.3.15) describes the behaviour of a derivative of the spectrum function at $\kappa=0$, and there must exist an equivalent condition for an integral of the correlation function $R_{ij}(\mathbf{r})$. We find from (5.3.1) and (2.4.3) that

$$2C_{ijpq}=\left[\frac{\partial^2\Phi_{ij}(\boldsymbol{\kappa})}{\partial\kappa_p\,\partial\kappa_q}\right]_{\kappa=0}=-\frac{1}{8\pi^3}\int r_p r_q R_{ij}(\mathbf{r})\,d\mathbf{r}, \qquad (5.3.16)$$

and the equivalent result is that the integral on the right side of (5.3.16) is invariant during decay. In the particular case of isotropic turbulence, for which $R_{ij}(\mathbf{r})$ has the form (3.4.5) and C_{ijlm} has the form (3.4.24), we readily find that the above result becomes

$$C=\frac{1}{3\pi}u^2\int_0^\infty r^4 f(r)\,dr=\text{constant during decay.} \qquad (5.3.17)$$

This condition on the function $f(r)$ was first obtained by Loitsiansky (1939), and the equivalent result for the isotropic spectrum function $E(\kappa)$ was pointed out by Lin (1947). The general result for homogeneous turbulence was put forward by Batchelor (1949*a*).

5.4. The final period of decay

We have seen that the energy spectrum at very small wave-numbers suffers very little modulation during the whole of the decay process. On the other hand, the energy in higher wave-numbers of the spectrum is being rapidly dissipated by viscosity, and it follows that ultimately the big eddies will supply most of the remaining energy of the turbulence. If we choose the current time t as any instant after this ultimate state has been reached, we have the opportunity of formulating a decay problem in which the initial form of the spectrum (or, rather, the relevant part of it) can be prescribed from the relation (5.3.1). This would not by itself make a tractable problem, but the assumption already made, that the decay is in an advanced stage, suggests that we might suppose with consistency that the turbulent velocities are so small as to make inertia forces negligible.† On this basis the dynamical equation is linear, and we are able to get a complete solution of the decay of the turbulence at very large times after its formation. It happens

† Giving a rather spurious kind of 'turbulence'.

that this final period of decay occurs at decay times which are within the reach of measurements in a wind-tunnel stream, and it has thus been possible to obtain valuable information about what, in the initial stages of decay, were the biggest eddies.

One of the two conditions defining the final period of decay is that t is so large that the non-linear terms (due to both inertia and pressure forces) in the Navier-Stokes equation are negligible, so that

$$\frac{\partial \mathbf{u}(\mathbf{x}, t)}{\partial t} = \nu \nabla^2 \mathbf{u}(\mathbf{x}, t). \tag{5.4.1}$$

This 'heat-conduction' equation may be solved in terms of an initial distribution of velocity in the realized field, as was suggested first by E. Reissner (1938). The corresponding equation for the Fourier coefficients is

$$\frac{\partial \, d\mathbf{Z}(\boldsymbol{\kappa}, t)}{\partial t} = -\nu\kappa^2 d\mathbf{Z}(\boldsymbol{\kappa}, t), \tag{5.4.2}$$

of which the solution is

$$d\mathbf{Z}(\boldsymbol{\kappa}, t) = d\mathbf{Z}(\boldsymbol{\kappa}, t_0)\, e^{-\nu\kappa^2(t-t_0)}, \tag{5.4.3}$$

where t_0 is an initial instant of time (which must, of course, lie within the period of validity of (5.4.1)). The corresponding variation of the spectrum tensor with time follows from (2.5.5) and (5.4.3) as

$$\Phi_{ij}(\boldsymbol{\kappa}, t) = \Phi_{ij}(\boldsymbol{\kappa}, t_0)\, e^{-2\nu\kappa^2(t-t_0)}, \tag{5.4.4}$$

so that the spectrum at an arbitrary time is determined when we know its initial form.

The exponential in (5.4.4) decreases rapidly as κ increases, and the other condition defining the final period of decay is that $(t-t_0)$ is so large that the right side of (5.4.4) is dominated by the first term in the expansion of $\Phi_{ij}(\boldsymbol{\kappa}, t_0)$ in powers of the components of $\boldsymbol{\kappa}$. That is, in view of (5.3.1),

$$\Phi_{ij}(\boldsymbol{\kappa}, t) \sim C_{ijlm}\kappa_l\kappa_m\, e^{-2\nu\kappa^2(t-t_0)} \tag{5.4.5}$$

for $(t-t_0)$ sufficiently large, where C_{ijlm} is independent of t according to the result established in the previous section. Thus, apart from the constant tensor C_{ijlm} which is determined by the initial conditions, the asymptotic form of the spectrum tensor is completely determined as a consequence of the relative permanence of the big eddies. It is difficult to make an *a priori* estimate of the decay time after which (5.4.5) will be valid, but since the variation of $e^{-2\nu\kappa^2(t-t_0)}$ with κ is very rapid the relation (5.4.5) is likely to hold soon after the first condition is satisfied, i.e. soon after inertia forces become negligible.

The velocity correlation tensor corresponding to the asymptotic spectrum tensor (5.4.5) is found from (2.4.2) to be

$$R_{ij}(\mathbf{r},t)=\frac{(2\pi^3)^{\frac{1}{2}}}{16[\nu(t-t_0)]^{\frac{5}{2}}}\left[C_{ijll}-\frac{r_l r_m C_{ijlm}}{4\nu(t-t_0)}\right]\exp\left[-\frac{r^2}{8\nu(t-t_0)}\right]. \tag{5.4.6}$$

In particular, the energy tensor in the final period is

$$\overline{u_i u_j}=\frac{(2\pi^3)^{\frac{1}{2}}}{16[\nu(t-t_0)]^{\frac{5}{2}}}C_{ijll},$$

and the mean squares of the velocity components are given by

$$\frac{\overline{u_1^2}}{C_{11ll}}=\frac{\overline{u_2^2}}{C_{22ll}}=\frac{\overline{u_3^2}}{C_{33ll}}=\frac{(2\pi^3)^{\frac{1}{2}}}{16[\nu(t-t_0)]^{\frac{5}{2}}}. \tag{5.4.7}$$

We have thus arrived at the interesting result that the energy of the turbulence ultimately decreases as the $(-\frac{5}{2})$ power of the time. Furthermore, the directional distribution of the energy in the asymptotic state is governed by the values of C_{ijll}, and is identical with the directional distribution of energy in what, in the initial stages of the decay, were the lowest wave-numbers of the spectrum. The asymptotic longitudinal correlation coefficient, in the direction of, say, the x_1-axis is

$$\frac{R_{11}(r,0,0,t)}{\overline{u_1^2}}=\exp\left[-\frac{r^2}{8\nu(t-t_0)}\right], \tag{5.4.8}$$

in view of the requirement of the continuity condition that $C_{1111}=0$ (see remark following (3.1.10)). This longitudinal correlation coefficient is the same for all directions of $\mathbf{r}$, and is also the same for all kinds of homogeneous turbulence.

These conclusions have been verified by direct measurements of the turbulence generated by a grid of bars lying in two perpendicular directions in a plane at right angles to a wind-tunnel stream (Batchelor and Townsend, 1948*b*). It was necessary to use a grid of small bar spacing ($M = 0{\cdot}16$ cm) and to use a small stream speed ($U = 620$–900 cm sec^{-1}) in order to achieve the small Reynolds numbers for which the linearized Navier-Stokes equation is applicable, and the final period of decay, as defined above, was found to set in at a distance of about $450M$ from the grid at the lowest speed used. At higher values of the grid Reynolds number (for $M = 0{\cdot}16$ cm and $U = 620$ cm sec^{-1}, the grid Reynolds number is

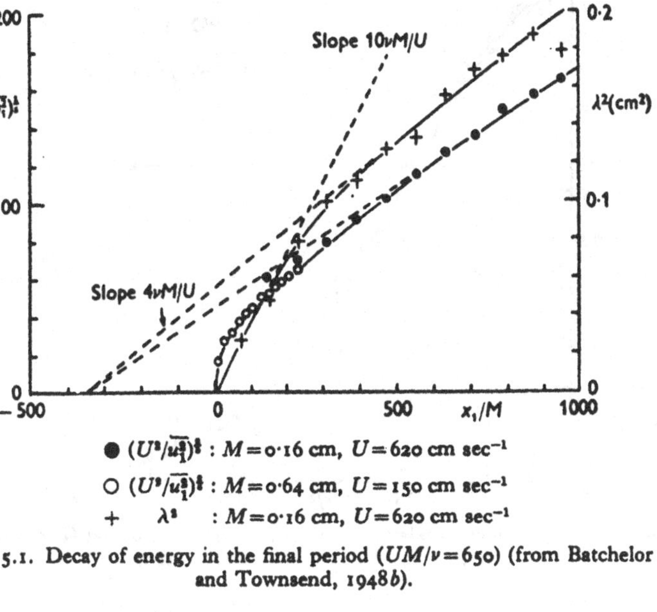

● $(U^2/\overline{u_1^2})^{\frac{1}{2}}$: $M = 0{\cdot}16$ cm, $U = 620$ cm sec^{-1}

○ $(U^2/\overline{u_1^2})^{\frac{1}{2}}$: $M = 0{\cdot}64$ cm, $U = 150$ cm sec^{-1}

+ λ^2 : $M = 0{\cdot}16$ cm, $U = 620$ cm sec^{-1}

Fig. 5.1. Decay of energy in the final period ($UM/\nu = 650$) (from Batchelor and Townsend, 1948*b*).

$UM/\nu = 650$) the final period of decay occurs at a much larger number of mesh lengths from the grid. Fig. 5.1 shows the measurements of $\overline{u_1^2}$, plotted in a form suited to the asymptotic decay law, and values of λ^2 calculated from measurements of $\overline{(\partial u_1/\partial x_1)^2}$, where

$$\frac{\overline{u_1^2}}{\lambda^2} = \overline{\left(\frac{\partial u_1}{\partial x_1}\right)^2} = -\left[\frac{\partial^2 R_{11}(r, 0, 0, t)}{\partial r^2}\right]_{r=0}$$

(the x_1-axis was chosen in the direction of the stream, x_1 being measured from the grid). The relation (5.4.8) shows that in the final period of decay λ^2 has the value $4\nu(t - t_0)$, and a straight line of slope 4ν (multiplied by a factor M/U to make the transition from the temporal decay of the theoretical turbulence to the spatial decay of the experimental turbulence) has been drawn in the figure for comparison with the measurements. Fig. 5.2 (in which λ has been taken as equal to the radius of curvature at $r=0$ of the measured curve in each case) shows the good agreement between the theoretical form (5.4.8) and measurements of the correlation coefficient at three stages of the decay.

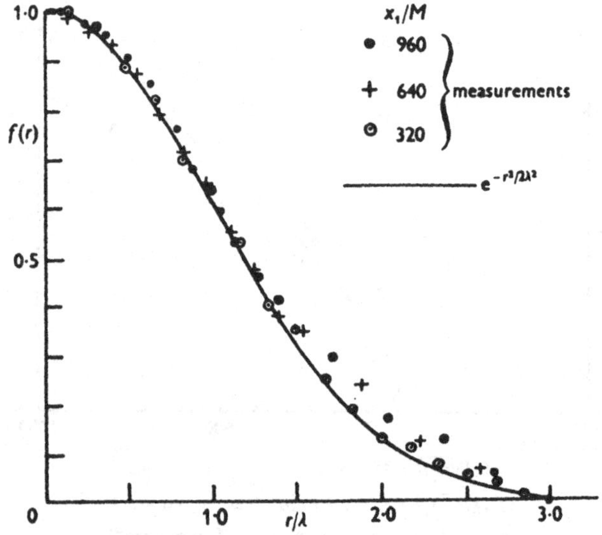

Fig. 5.2. Longitudinal correlation coefficient in the final period ($UM/\nu=650$) (from Batchelor and Townsend, 1948*b*).

Fig. 5.3 shows measurements of the ratio $\overline{u_1^2}/\overline{u_2^2}$ ($\overline{u_1^2}/\overline{u_3^2}$ had effectively the same values) in both the initial and final stages of the decay (Batchelor and Stewart, 1950). After being close to unity in the initial stages of the decay (when the turbulence is isotropic, so far as the bulk of the energy is concerned), $\overline{u_1^2}/\overline{u_2^2}$ increases up to an asymptotic value in the neighbourhood of 1·5. The turbulence is anisotropic in the final period of decay after being apparently isotropic earlier, which implies that the motion associated with the smallest wave-numbers at the instant of formation of the homogeneous turbulence is not isotropic. This cannot be regarded as surprising† in view of the very marked directional character of the grid and the method of generating the turbulence. Presumably the row of bars parallel to the x_2-axis produces large contributions to $\overline{u_1^2}$ and $\overline{u_3^2}$, while the row parallel to the x_3-axis produces large contributions to $\overline{u_1^2}$ and $\overline{u_2^2}$, the net result being that the contribution to $\overline{u_1^2}$ from the smallest wave-numbers is permanently greater than the contributions to $\overline{u_2^2}$ or $\overline{u_3^2}$.

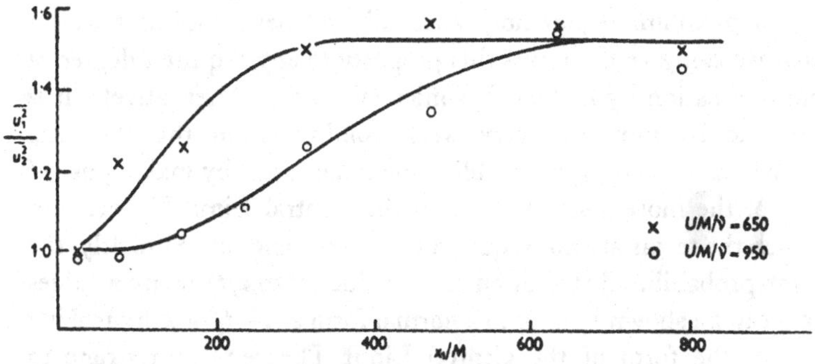

Fig. 5.3. Directional distribution of energy in the final period (after Batchelor and Stewart, 1950).

A prediction about the velocity field in the final period of decay which has not yet been tested experimentally can be obtained with the aid of the Central Limit Theorem. The relation between the velocity fields at times t_0 and t follows from equation (5.4.1) as

$$\mathbf{u}(\mathbf{x}, t) = \frac{1}{[4\pi\nu(t-t_0)]^{\frac{3}{2}}} \int \mathbf{u}(\mathbf{x}', t_0) \exp\left[-\frac{(\mathbf{x}-\mathbf{x}')^2}{4\nu(t-t_0)}\right] d\mathbf{x}'. \tag{5.4.9}$$

As $t-t_0$ tends to infinity, the exponential factor in this integral becomes different from zero (and approaches unity) for a larger and larger region of $\mathbf{x}'$-space centred on the point $\mathbf{x}$. Hence if we regard the integral as being equivalent to the sum of a number of sub-integrals, each extending over a volume V (say) of $\mathbf{x}'$-space, the effect of increasing $t-t_0$ is to increase the number of terms in the expression for $\mathbf{u}(\mathbf{x}, t)$. Each of the terms in this expression is

† If we were not so accustomed to the phenomenon, we might be surprised that the turbulence behind a grid is so nearly isotropic in the initial stages of the decay.

a random variable with finite second moment, so that we have here a type of problem to which the Central Limit Theorem is applicable. In its simplest form, the theorem says† that the probability distribution of the sum (i.e. of $\mathbf{u}(\mathbf{x}, t)$) tends to a normal or Gaussian form as the number of terms tends to infinity, provided the various terms are statistically independent. In the above case the terms, each of them being essentially the integral $\int \mathbf{u}(\mathbf{x}', t_0)\, d\mathbf{x}'$ taken over a certain region of volume V, are not statistically independent, but it is clear that we can give the terms this property to any required degree of approximation by making V sufficiently large. Alternatively, it is possible by imposing very weak conditions on the statistical properties of $\mathbf{u}(\mathbf{x}', t_0)$ to establish the same result by making use of one of the more general forms of the Central Limit Theorem in which the terms are not required to be independent. Similarly, the joint-probability distribution of the values of $\mathbf{u}(\mathbf{x}, t)$ at any n values of $\mathbf{x}$ can be shown to tend to a normal form as $t - t_0 \to \infty$ by making use of the form of the Central Limit Theorem appropriate to a $3n$-dimensional variable. (Cramér, op. cit. p. 113, gives a statement of the theorem for the case of a sum of independent $3n$-dimensional terms.)

Hence, in the final period of decay, the statistical distribution of the velocity field tends asymptotically to the normal form, the second moment being given by the expression (5.4.6). The final period of decay provides one case in which we are able to deduce (the limiting form of) the complete statistical representation of the turbulence at one instant, as well as a complete description of one realization of the velocity field at time t in terms of the same field at time t_0.

It is not difficult to obtain also the asymptotic form of the mean product of the velocities at two different times of decay (Batchelor and Townsend, 1948*b*); as would be expected, correlation between the velocities persists over a longer time interval at larger times of decay.

† See H. Cramér, *Random variables and probability distributions*, Cambridge University Press, 1937.

5.5. Dynamical equations for isotropic turbulence

Since we shall be concerned primarily with isotropic turbulence—or rather, with that part of the spectrum which becomes isotropic under the action of pressure forces—in the remaining chapters, we shall set out here the formal dynamical equations for this case. We have seen in § 3.4 that the condition of isotropy fully determines the dependence of the spectrum tensor $\Phi_{ij}(\boldsymbol{\kappa}, t)$ on the direction of $\boldsymbol{\kappa}$, and a discussion of the dynamics of isotropic turbulence is therefore concerned wholly with the exchange of energy between wave-numbers of different magnitude. This permission to think of the effect of eddy size alone greatly facilitates the construction of physical hypotheses about the transfer of energy, as will be seen in the following chapters.

When the turbulence is isotropic, the tensor equations (5.2.10) and (5.2.11) each reduce to a single scalar equation. Taking first the dynamical equation for $R_{ij}(\mathbf{r})$, we have found (see equations (3.4.5) and (3.4.6)) that $R_{ij}(\mathbf{r})$ can be expressed in terms of the scalar function $u^2 f(r)$ as

$$R_{ij}(\mathbf{r}) = u^2 \left[-\frac{1}{2r} f' r_i r_j + (f + \tfrac{1}{2} r f') \delta_{ij} \right],$$

where $f' = \partial f / \partial r$. It was also established that the solenoidal first-order tensor $\overline{p u_i'}$ is identically zero in isotropic turbulence, so that the term $P_{ij}(\mathbf{r})$ is zero (as we expect from the interpretation of the effect of $P_{ij}(\mathbf{r})$ as a tendency to isotropy). Denoting the triple velocity correlation $\overline{u_i u_l u_j'}$ by $S_{ilj}(\mathbf{r})$, the term $T_{ij}(\mathbf{r})$ is (see (5.2.12))

$$T_{ij}(\mathbf{r}) = \frac{\partial S_{ilj}(\mathbf{r})}{\partial r_l} - \frac{\partial S_{jli}(-\mathbf{r})}{\partial r_l}, \tag{5.5.1}$$

and we found in § 3.4 that $S_{ilj}(\mathbf{r})$ can be expressed in terms of the scalar function $u^3 k(r)$ (see (3.4.32)). Consequently, (5.2.10) is in effect an equation for $\frac{\partial}{\partial t} u^2 f(r)$ in terms of $u^3 k(r)$, and derivatives of $f(r)$ arising from the viscosity term. The easiest method of obtaining this scalar equation is to put $i = j$ in (5.2.10) and sum over all values of i. We have

$$R_{ii}(\mathbf{r}) = u^2 (3f + rf') = 2R(r), \tag{5.5.2}$$

as already displayed in (3.4.14), and

$$\nabla^2 R_{ii}(\mathbf{r}) = 2\left(\frac{\partial^2}{\partial r^2} + \frac{2}{r}\frac{\partial}{\partial r}\right) R(r);$$

also, in view of (3.4.35),

$$T_{ii}(\mathbf{r}) = \frac{\partial}{\partial r_l}[S_{ili}(\mathbf{r}) - S_{ili}(-\mathbf{r})]$$

$$= \left(r\frac{\partial}{\partial r} + 3\right) K(r), \tag{5.5.3}$$

where $K(r)$ stands for $u^3\left(\frac{\partial}{\partial r} + \frac{4}{r}\right) k(r)$. Hence for isotropic turbulence (5.2.10) reduces to

$$\frac{\partial R(r)}{\partial t} = \frac{1}{2}\left(r\frac{\partial}{\partial r} + 3\right) K(r) + 2\nu\left(\frac{\partial^2}{\partial r^2} + \frac{2}{r}\frac{\partial}{\partial r}\right) R(r). \tag{5.5.4}$$

A first integral of this equation is

$$\frac{\partial u^2 f(r)}{\partial t} = u^3\left(\frac{\partial}{\partial r} + \frac{4}{r}\right) k(r) + 2\nu u^2\left(\frac{\partial^2}{\partial r^2} + \frac{4}{r}\frac{\partial}{\partial r}\right) f(r), \tag{5.5.5}$$

which is the dynamical equation derived first by v. Kármán and Howarth (1938); note that although not suggested by the notation, the quantities u, f and k in this equation are functions of the time of decay.

Putting $r=0$ in (5.5.5) gives the rate of change of energy as

$$\frac{d\frac{3}{2}u^2}{dt} = 15\nu u^2 f_0'' = -\frac{15\nu u^2}{\lambda^2}; \tag{5.5.6}$$

the proportional rate of decrease of energy is

$$\frac{1}{u^2}\frac{du^2}{dt} = -\frac{10\nu}{\lambda^2},$$

whence comes the term 'dissipation† length parameter' for λ. We note the interesting result, for the special case of a power-law decay of kinetic energy, $u^2 \propto t^{-n}$, that

$$\lambda^2 = \frac{10\nu t}{n}. \tag{5.5.7}$$

Equating the coefficients of other powers of r^2 in the expansion of

† But note that λ is not the length scale of the eddies responsible for most of the dissipation (see § 6.4).

all terms of (5.5.5) yields the rate of change of other physical quantities. For the coefficients of r^2 we find the equation

$$\frac{\mathrm{d}}{\mathrm{d}t}(\tfrac{1}{2}u^2 f_0'') = \tfrac{7}{6}u^3 k_0''' + \tfrac{7}{3}\nu u^2 f_0^{\mathrm{iv}}, \tag{5.5.8}$$

which, in view of (3.4.8), is the equation for the rate of change of mean-square vorticity.

The corresponding equation (5.2.11) for the spectrum tensor similarly reduces to a single scalar equation. Again the scalar equation emerges most easily if we contract the indices to obtain

$$\frac{\partial}{\partial t}\Phi_{ii}(\boldsymbol{\kappa}) = \Gamma_{ii}(\boldsymbol{\kappa}) - 2\nu\kappa^2\Phi_{ii}(\boldsymbol{\kappa}), \tag{5.5.9}$$

where, according to (3.4.13),

$$\Phi_{ii}(\boldsymbol{\kappa}) = \frac{E(\kappa)}{2\pi\kappa^2},$$

and, according to (5.3.3) and (3.4.34),

$$\Gamma_{ii}(\boldsymbol{\kappa}) = 2\kappa^4\Upsilon(\boldsymbol{\kappa}). \tag{5.5.10}$$

Consequently the dynamical equation for the energy-spectrum function $E(\kappa)$ is

$$\frac{\partial E(\kappa)}{\partial t} = T(\kappa) - 2\nu\kappa^2 E(\kappa), \tag{5.5.11}$$

where $T(\kappa)$ represents the contribution due to transfer of energy from all other wave-numbers and is related to $\Upsilon(\kappa)$ by

$$T(\kappa) = 4\pi\kappa^6\Upsilon(\kappa). \tag{5.5.12}$$

The connexion between (5.5.4) or (5.5.5) and (5.5.11) is established from the Fourier transform relations (see (3.4.15), (3.4.36) and (3.4.37))

$$\begin{Bmatrix} E(\kappa) \\ T(\kappa) \end{Bmatrix} = \frac{2}{\pi}\int_0^\infty \begin{Bmatrix} R(r) \\ \dfrac{1}{2}\left(r\dfrac{\partial}{\partial r}+3\right)K(r) \end{Bmatrix} \kappa r \sin \kappa r \,\mathrm{d}r, \tag{5.5.13}$$

$$\begin{Bmatrix} R(r) \\ \dfrac{1}{2}\left(r\dfrac{\partial}{\partial r}+3\right)K(r) \end{Bmatrix} = \int_0^\infty \begin{Bmatrix} E(\kappa) \\ T(\kappa) \end{Bmatrix} \frac{\sin \kappa r}{\kappa r}\,\mathrm{d}\kappa. \tag{5.5.14}$$

These formal relations between double- and triple-velocity correlations and their transforms slightly obscure the way in which the Fourier coefficients $\mathrm{d}\mathbf{Z}(\boldsymbol{\kappa})$ interact to produce the transfer of energy denoted by $T(\kappa)$ in (5.5.11). We can recover this more fundamental information from (5.2.6). Whether the turbulence is isotropic or not,

$$\frac{\partial E(\kappa)}{\partial t}=\frac{\partial}{\partial t}\frac{1}{2}\int \Phi_{ii}(\boldsymbol{\kappa})\,dA(\kappa)$$
$$=\iint Q(\boldsymbol{\kappa},\boldsymbol{\kappa}')\,d\boldsymbol{\kappa}'\,dA(\kappa)-2\nu\kappa^2E(\kappa), \qquad (5.5.15)$$

where $dA(\kappa)$ is an element of area of a circle of radius κ centred at the origin, and $Q(\boldsymbol{\kappa},\boldsymbol{\kappa}')$ is given in terms of the Fourier coefficients by (5.2.7). The rate of change of energy associated with all vector wave-numbers whose magnitudes are less than κ is

$$\frac{\partial\int_0^\kappa E(\kappa'')\,d\kappa''}{\partial t}=\iint_{(\kappa''<\kappa)} Q(\boldsymbol{\kappa}'',\boldsymbol{\kappa}')\,d\boldsymbol{\kappa}'\,d\boldsymbol{\kappa}''-2\nu\int_0^\kappa \kappa''^2E(\kappa'')\,d\kappa''$$
$$=-\int_{(\kappa'>\kappa)}\int_{(\kappa''<\kappa)} Q(\boldsymbol{\kappa}',\boldsymbol{\kappa}'')\,d\boldsymbol{\kappa}'\,d\boldsymbol{\kappa}''$$
$$-2\nu\int_0^\kappa \kappa''^2E(\kappa'')\,d\kappa'', \qquad (5.5.16)$$

in view of the antisymmetry of $Q(\boldsymbol{\kappa}',\boldsymbol{\kappa}'')$ in $\boldsymbol{\kappa}''$ and $\boldsymbol{\kappa}'$ (see (5.2.8)). When the turbulence is isotropic, $Q(\boldsymbol{\kappa}',\boldsymbol{\kappa}'')$ must be a function of κ', κ'', $\boldsymbol{\kappa}'.\boldsymbol{\kappa}''$ (and t) alone, so that

$$\frac{\partial}{\partial t}\int_0^\kappa E(\kappa'')\,d\kappa''$$
$$=-\int_{\kappa'=\kappa}^{\infty}\int_{\theta'=0}^{\pi}\int_{\kappa''=0}^{\kappa}\int_{\theta''=0}^{\pi} Q(\kappa',\kappa'',\kappa'\kappa''\cos(\theta'-\theta''))$$
$$\times\, d\kappa'\,d(\cos\theta')\,d\kappa''\,d(\cos\theta'')-2\nu\int_0^\kappa \kappa''^2E(\kappa'')\,d\kappa''$$
$$=-\int_{\kappa'=\kappa}^{\infty}\int_{\kappa''=0}^{\kappa} P(\kappa',\kappa'')\,d\kappa'\,d\kappa''-2\nu\int_0^\kappa \kappa''^2E(\kappa'')\,d\kappa'', \qquad (5.5.17)$$

where

$$P(\kappa',\kappa'')=\int_{-1}^{1}\int_{-1}^{1} Q(\kappa',\kappa'',\kappa'\kappa''\cos(\theta'-\theta''))\,d(\cos\theta')\,d(\cos\theta''). \qquad (5.5.18)$$

Also, in the case of isotropic turbulence, a comparison of (5.2.6) and (5.5.9) shows that

$$\int Q(\boldsymbol{\kappa},\boldsymbol{\kappa}')\,d\boldsymbol{\kappa}'=\tfrac{1}{2}\Gamma_{ii}(\boldsymbol{\kappa})=\kappa^4\Upsilon(\kappa)=\frac{1}{4\pi\kappa^2}T(\kappa). \qquad (5.5.19)$$

CHAPTER VI

THE UNIVERSAL EQUILIBRIUM THEORY

6.1. The hypothesis of statistical equilibrium

It has long been the practice in statistical mechanics to seek out problems in which the statistical conditions are uniform, in view of their much greater tractability. It is very natural therefore to make the same search in the field of turbulence. The problem has already been specialized in this book to the case of turbulence which is spatially uniform; are there now any aspects or parts of homogeneous turbulence on which the temporal decay does not have an appreciable direct effect? The question may also be put in mechanical terms which make clear the basis for the steadiness: are there any degrees of freedom of the dynamical system for which the forces are in approximate statistical equilibrium?

It will be useful to consider first the time scale of the decay of the total kinetic energy in order to see where we may *not* expect to find statistical equilibrium. If we can regard the range of wave-numbers containing most of the energy as forming a definite group (the 'energy-containing eddies'), with characteristic velocity u $(=(\frac{1}{3}\overline{\mathbf{u}^2})^{\frac{1}{2}})$ and characteristic length l (which might be, for instance, the reciprocal of the wave-number at which the maximum of the energy spectrum function $E(\kappa, t)$ occurs), it is permissible to think also of a characteristic time l/u of the energy-containing eddies. We want to compare this characteristic time with the time scale of the decay of energy in order to see how close to—or far from—equilibrium are the energy-containing eddies. Now it has been remarked before that the rate of decrease of kinetic energy is found experimentally to be of the order of u^3/l (some of the evidence is described in the appendix to this section) during an initial period of the decay in which inertia forces are appreciable; that is,

$$\frac{du^2}{dt} = -\frac{Au^3}{l}, \tag{6.1.1}$$

where A is a number of order unity (which may vary slightly with the time of decay and the initial conditions of the turbulence and the choice of l). The time scale of the decay of energy is

$$u^2 \bigg/ \left|\frac{du^2}{dt}\right| = \frac{1}{A}\frac{l}{u},$$

and is thus of the order of the characteristic time of the energy-containing eddies. It appears that the whole life (or, at any rate, that part of it for which (6.1.1) is valid) of the turbulence is not longer in duration than several characteristic periods of the energy-containing eddies, the reason being that the process of adjustment of the energy-containing eddies *is* the process of decay. We have here a situation which is very far from that of the kinetic theory of gases, and the energy-containing eddies are far from being in approximate equilibrium under the action of inertia and viscous forces.

However, this does not exhaust the possibilities. We saw in the previous chapter that inertia forces result in the spreading of the energy of the turbulence over a wider and wider range of wave-numbers, and that this process will be checked only by the stronger viscous damping at large wave-numbers. We may expect that the spectrum function $E(\kappa, t)$ will have a maximum at some value of κ and will fall off to zero monotonically as $\kappa \to \infty$. Provided that the characteristic time of eddies decreases as their size decreases —which is intuitively very plausible and is in accord with visual observations of turbulent motion, but will need to be checked *a posteriori*—there is a chance that some parts of the spectrum will have a characteristic time small compared with the time scale of the over-all decay and will therefore be associated with degrees of freedom which are in approximate statistical equilibrium. The concept of a characteristic time of eddies of a certain size is so indefinite that we cannot do more than suggest the existence, at sufficiently large wave-numbers and under suitable conditions, of a range of the spectrum belonging to degrees of freedom in equilibrium, and rely on experiments to confirm or contradict the suggestion. So far as they go, the available measurements are consistent with the hypothesis, and as a consequence the idea of an equilibrium range of the spectrum, and the theory which has been built upon it, constitutes the most important development of the last ten years. This development, to which the present chapter is devoted, is the theory put forward first by A. N. Kolmogoroff (1941*a*, *c*), suggested independently by L. Onsager (1945, 1949) and C. F. von Weizsäcker (1948) some years later, and partly anticipated by A. M. Obukhoff (1941).

Appendix to § 6.1. The empirical relation (6.1.1) is so useful and relevant to the ideas of the present chapter that some of the evidence will be presented here, even though the more detailed discussion of the energy-containing eddies must wait until Chapter VII. A choice of the length of l which permits direct measurements is the longitudinal integral scale

$$L_p(t) = \frac{\int_0^\infty R_{11}(r, 0, 0, t)\, dr}{R_{11}(0, 0, 0, t)},$$

which, for isotropic turbulence, in view of (3.4.21), is related to the spectrum function in the following way:

$$L_p(t) = \int_0^\infty f(r, t)\, dr = \frac{3\pi}{4} \frac{\int_0^\infty \kappa^{-1} E(\kappa, t)\, d\kappa}{\int_0^\infty E(\kappa, t)\, d\kappa}. \qquad (6.1.2)$$

This length $L_p(t)$ is not as directly representative of the part of the function $E(\kappa, t)$ that makes the major contribution to the total energy $\int_0^\infty E(\kappa, t)\, d\kappa$ as it might be, for it gives too much weight to small values of κ, but the difference is not enough to obscure the point at issue. Measurements of L_p at various stages of decay and at several values of the Reynolds number UM/ν of the grid producing the turbulence have been reported (Batchelor and Townsend, 1948*a*), together with measurements of u and λ made under the same conditions. The measurements can therefore be used to determine

$$-\frac{L_p}{u^3}\frac{du^2}{dt} = 10 \frac{L_p}{\lambda} \frac{\nu}{u\lambda},$$

which, according to (6.1.1), should be approximately constant and of order unity. Fig. 6.1 shows that the values of this quantity obtained from the measurements are in fair agreement with (6.1.1). Scattered observations of L_p, u and λ at other grid Reynolds numbers are also consistent with (6.1.1). R. W. Stewart (1951) has shown experimentally that the part of the correlation function $f(r, t)$ that is determined by the energy-containing eddies has approximately the same shape at various stages of the decay when plotted as a function of r/L_p, for $UM/\nu = 5300$; this shape (with a suitable transformation of the ordinate scale, see §7.2), is also approximately independent of Reynolds number, suggesting that the values of l and u at any moment are sufficient to determine the instantaneous conditions and in particular the approximate rate of decay, which is consistent with (6.1.1). Other relevant evidence is presented by H. L. Dryden (1943).

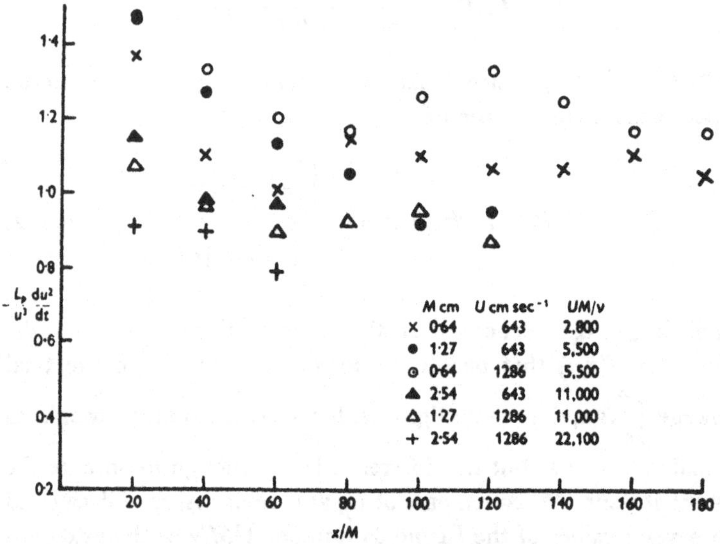

Fig. 6.1. The empirical relation $\dfrac{du^2}{dt} = -A\dfrac{u^3}{l}$.

6.2. Turbulent motion at large Reynolds number

The consequence of the hypothesis of statistical equilibrium at sufficiently high wave-numbers is that the rates of change of mean values (and their transforms) determined by this equilibrium range of wave-numbers can be regarded as negligible. Although a useful simplification, this hypothesis by itself does not permit definite predictions to be made. It is still necessary to know something about the transfer of energy between different wave-numbers and about the nature of the equilibrium. An additional hypothesis which fills this need will be described in the next section, and in order that it will appear natural, we shall first put forward some general ideas about turbulent motion at large Reynolds number.

We have seen that inertia forces tend to spread energy over as wide a range of wave-numbers as is consistent with the stronger viscous damping which occurs at large wave-numbers. The balance produced by these two opposing tendencies is controlled by the Reynolds number of the turbulence (we can leave aside, for the moment, the question of how to choose the length and velocity which determines the Reynolds number), and the higher the Reynolds number, i.e. the smaller the value of ν, the greater will be the success of the inertia forces in transferring energy to higher and higher wave-numbers. One may regard each degree of freedom of the motion (i.e. each wave-number component) as having its own Reynolds number—a possible definition would be $[E(\kappa, t)]^{\frac{1}{2}}/\nu\kappa^{\frac{1}{2}}$—and the effect of decreasing ν is to increase the dominance of the inertia forces over viscous forces for the motion associated with that degree of freedom. Consequently the region of wave-number space which is affected significantly by the action of viscous forces moves out from the origin towards $\kappa = \infty$ as the Reynolds number increases. In the limit of infinite Reynolds number the sink of energy is displaced to infinity and the influence of viscous forces is negligible for wave-numbers of finite magnitude.

This dominance of the motion by inertia forces when the Reynolds number is large, for all except components of very large wave-number, occurs in other hydrodynamical situations. When fluid flows steadily past a rigid boundary, the effect of viscous forces is confined, approximately, to a boundary layer which decreases in thickness as the Reynolds number of the flow increases. At very large Reynolds numbers viscous forces are effective only within this layer in which the velocity changes rapidly from zero at the boundary to the appropriate value (determined by inertia forces and the geometry of the boundary) at the outer edge of the boundary layer. If a Fourier analysis of the velocity distribution were made, the near-discontinuity of the velocity at the boundary would be represented by the very large wave-number components, so that here too the effect of viscous forces is negligible, except in the outer parts of wave-number space, when the Reynolds number is large. Mathematically the process can readily be understood, in a qualitative way, as a balance between the tendency to form discontinuities, arising from the non-linear term $\mathbf{u}.\nabla\mathbf{u}$ in the equation of motion, and the damping term $\nu\nabla^2\mathbf{u}$; as the parameter ν approaches zero, the steep gradients, formed by the non-linear term and permitted by the damping term, become closer to discontinuities. The existence of a rigid boundary, with a no-slip condition, places these layers of rapid change in velocity at the boundary, but in turbulent flow they occur in the interior of the fluid. The existence of dissipation layers in turbulent motion has been emphasized by J. M. Burgers (1948*a*) who has investigated them quantitatively for a hypothetical simplified equation of motion.† More will be said about this important matter in Chapter VIII.

The relegation of the influence of viscous forces to the outer parts of wave-number space at large Reynolds numbers may be interpreted in another way, suggested by the known results concerning the stability of hydrodynamical motions. There are some steady flow fields which are unstable to small periodic disturbances of wave-number κ_1, say, for Reynolds numbers greater than a certain critical value R_1, and unstable to other disturbances of wave-numbers $\kappa_2, \kappa_3, \ldots$ for Reynolds numbers above $R_2, R_3, \ldots$ respectively, where $R_1 < R_2 < R_3 \ldots$. (An example is the steady flow between two concentric rotating cylinders when the circulation about the inner cylinder is greater than the circulation about the outer cylinder.‡) Hence if the Reynolds number of such a flow field is increased from zero to infinity, the various instabilities enter (or would enter if the same steady flow could be preserved at all Reynolds numbers) at the Reynolds number for which the inertia forces are just dominant (in the sense of being able to transfer energy to the disturbance more rapidly than viscous forces are dissipating the energy of the disturbance) for a disturbance of certain wave-number. We can regard the entrance of an instability as being the excitation of one more degree of freedom (or normal mode of

† J. D. Cole (*Quart. Appl. Math.* **9**, 1951, 225) has found that solutions of the equation $\frac{\partial u}{\partial t} + u\frac{\partial u}{\partial x} = \nu\frac{\partial^2 u}{\partial x^2}$ have the same general properties.

‡ See D. Meksyn, 'Stability of flow between rotating cylinders', *Proc. Roy. Soc.* A, **187**, 1948, 115.

motion). The number of degrees of freedom which are excited at any given stage depends on the Reynolds number, and at infinite Reynolds number all those modes of motion of which the system is capable are excited. In the case of turbulent motion, there is a continuum of degrees of freedom, all of which are excited to some extent, so that Reynolds numbers are not 'critical' in the stability sense, but it is still true that raising the Reynolds number permits the domination of more degrees of freedom by inertia forces and reduces the number which are (relatively) suppressed by viscous damping. We may anticipate that at finite (but large) Reynolds number the spectral density of energy in turbulent flow will begin to decrease sharply at some large wave-number which marks the beginning of the range of wave-numbers at which viscous forces are not negligible.

6.3. The hypothesis of independence of Fourier components for distant wave-numbers

The picture of turbulent motion at large Reynolds numbers which has been described can now be made the basis of an hypothesis which supplements and greatly increases the usefulness of the hypothesis of statistical equilibrium at large wave-numbers.

In the early stages of the generation of a field of turbulence, of whatever kind, only the smaller wave-numbers of the eventual spectral distribution of energy are excited. These smaller wave-numbers are those which are of the order of magnitude of the reciprocal of the various linear dimensions of the mechanical system generating the turbulence (for example, the diameters of and the distance between the bars of a grid placed across a stream, or the diameter of a pipe through which fluid is being forced under pressure), and they receive energy directly from that mechanical system. Then, as we have seen, the action of inertia forces is to transfer energy to other (and, in general, higher) wave-numbers and to direct it to the sink provided by viscous dissipation. There will thus be a range of wave-numbers which is not excited directly by the external large-scale forces which generate the motion, and which owes its excitation entirely to the energy transfer by inertia forces. Although we have no direct information about the nature of the transfer process, it is very plausible that the influence of the external conditions is strongest for the small wave-numbers on which they act directly, is less strong for the adjacent range of wave-numbers, and disappears altogether for those high wave-numbers which are at the end of a long chain of inertial transfer processes. The process of statistical transfer of energy across the spectrum will surely be accompanied by a loss of the order or 'information' contained in this energy.

If it is true that the process of transfer of energy to higher wave-numbers is accompanied by a weakening of the influence of the large-scale conditions of the motion, an immediate prediction follows from our conclusion in Chapter v that the effect of pressure forces is to tend to eliminate directional preferences in the energy associated with each volume element of wave-number space; whatever the directional preferences of the large-scale components of the motion, the motion associated with sufficiently large wave-numbers should be isotropic. This is a prediction which is readily tested experimentally, and has now received considerable support (Townsend, 1948*b*; Corrsin, 1949; Corrsin and Uberoi, 1950; Laufer, 1950). Many different tests of isotropy of the small-scale components may be used, and the success of the above prediction has been found to vary with the test adopted. Townsend (1948*b*) checked that the relations

$$\overline{\left(\frac{\partial u_1}{\partial x_1}\right)^2} = \frac{1}{2}\overline{\left(\frac{\partial u_2}{\partial x_1}\right)^2} = \frac{1}{2}\overline{\left(\frac{\partial u_3}{\partial x_1}\right)^2}, \qquad (6.3.1)$$

which are valid for isotropic turbulence (see (3.4.7)), are approximately valid at different positions across the turbulent wake of a cylinder (the x_1-axis is in the direction of the stream). The mean square quantities in (6.3.1) weight the small-scale components as strongly as does the expression for the dissipation, so that if the Reynolds number of the experiment is high enough to push the dissipation into wave-numbers which are out of reach of the very definite directional influence of the large-scale components (which is revealed by the fact that $\overline{u_1 u_2}$ was found to be about $0{\cdot}5\overline{u_1^2}$), the relations (6.3.1) should hold. In fact, the experiments showed that these relations were satisfied at Reynolds numbers smaller than that required to separate the energy-containing and dissipation ranges, so that the theoretical prediction is evidently here being aided by some other effect. A sensitive test of isotropy of the small-scale components (due to Corrsin) has been used by the other authors referred to, and the measurements made by Laufer (1950) in the turbulent flow under pressure between parallel planes can be quoted as typical. In effect Laufer compared measurements of two components of the one-dimensional spectrum tensor (defined as in (3.1.1)), viz. $\Theta_{11}(\kappa_1)$ and $\Theta_{12}(\kappa_1)$, where the x_1-axis is in the direction of flow and the x_2-axis is perpendicuiar to the planes. As shown in fig. 6.2, the value of $\Theta_{12}(\kappa_1)$ is definitely non-zero at small

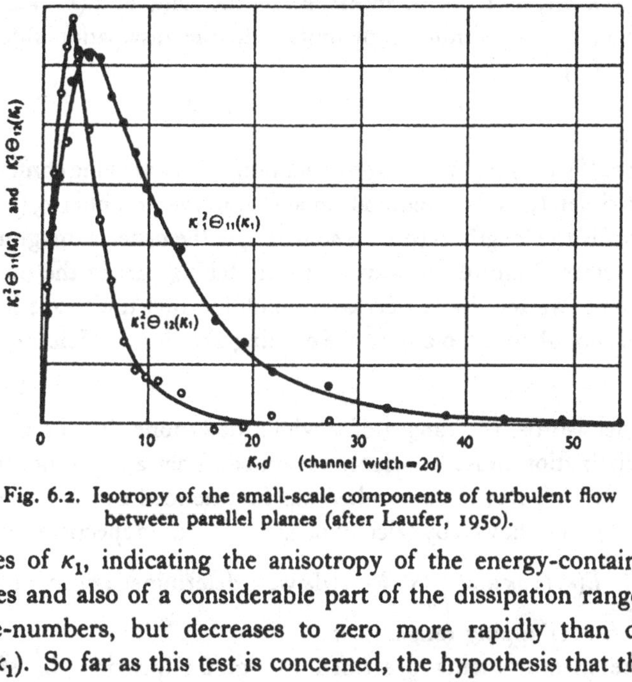

Fig. 6.2. Isotropy of the small-scale components of turbulent flow between parallel planes (after Laufer, 1950).

values of κ_1, indicating the anisotropy of the energy-containing eddies and also of a considerable part of the dissipation range of wave-numbers, but decreases to zero more rapidly than does $\Theta_{11}(\kappa_1)$. So far as this test is concerned, the hypothesis that there exists a range of (high) wave-numbers for which the associated motion is isotropic is permissible; and the same conclusion holds for the other turbulent flows that have been investigated.

With this encouragement we can proceed to formulate more definitely the idea of partial statistical independence of the high wave-number components of the motion.† Instead of referring to

† Called the 'disorder hypothesis' by C. F. von Weizsäcker (1948), and regarded as a consequence of the 'cascade process' of energy transfer by L. Onsager (1945).

the motion associated with the degree of freedom represented by the wave-number $\boldsymbol{\kappa}$, we can say simply, the Fourier coefficient $d\mathbf{Z}(\boldsymbol{\kappa}, t)$. The hypothesis is then as follows:

'At Reynolds numbers and for values of $\boldsymbol{\kappa}$ such that the Fourier coefficients $d\mathbf{Z}(\boldsymbol{\kappa}, t)$ are determined principally by the non-linear inertia term in the dynamical equation (5.2.4), and not by the viscous damping term, $d\mathbf{Z}(\boldsymbol{\kappa}, t)$ is statistically independent of $d\mathbf{Z}(\boldsymbol{\kappa}', t)$ if $|\boldsymbol{\kappa}| \gg |\boldsymbol{\kappa}'|$ or $|\boldsymbol{\kappa}| \ll |\boldsymbol{\kappa}'|$.'

The wave-number magnitude at which viscous forces do play an important part can be measured by the value of κ ($\kappa = \kappa_d$, say) at which the maximum contribution to the dissipation integral (see (5.2.9))

$$2\nu \int_0^\infty \kappa^2 E(\kappa, t)\, d\kappa, \quad = \epsilon, \text{ say}, \tag{6.3.2}$$

occurs. If we regard the energy-containing eddies as being confined† approximately to the neighbourhood of the wave-number κ_0 ($= 1/l$, where l is the length in (6.1.1)) so that κ_0 is the order of magnitude of the lower limit of the wave-numbers taking part in the inertial exchange, we see that a necessary condition that the above hypothesis should apply to any of the existing Fourier coefficients is

$$\kappa_0 \ll \kappa_d; \tag{6.3.3}$$

in other words, the ranges of κ which determine the energy and the dissipation must be widely separated. That a separation does occur at high Reynolds number was first noticed by G. I. Taylor (1938*b*), who showed by calculation from measured spectrum curves that $\int_0^\infty E(\kappa, t)\, d\kappa$ and $\int_0^\infty \kappa^2 E(\kappa, t)\, d\kappa$ were determined approximately by non-overlapping ranges of κ.

The measurements described in fig. 6.1 provide, in effect, evidence consistent with the above hypothesis. The measurements showed that for similar initial conditions of the turbulence, but for different decay times and different Reynolds numbers, the relation (6.1.1) is valid. We may think of the different Reynolds numbers as being produced by varying ν and keeping l and u constant, in which

† Spectrum curves derived from measurements of the turbulence generated by a square-mesh grid show a fairly dominant peak at $\kappa = 1/\alpha M$, where M is the spacing of the bars and α is a number, of order unity, which increases during the decay.

case the result says that the rate of decay of energy is independent of the viscosity. The rate of decay is identical with the rate at which energy is transferred, by the action of inertia forces, from the energy-containing range of wave-numbers to higher wave-numbers, so that we have here a demonstration that changes in ν, which will be accompanied by changes in the motion associated with the dissipation range of wave-numbers, have no effect on the rate of transfer of energy from the lower wave-numbers. If we write (6.1.1) in the form

$$\frac{\mathrm{d}u^2}{\mathrm{d}t} = -Au^2\frac{u}{l},$$

it again suggests (remembering that A is of order unity) that energy transfer occurs chiefly as a result of inertial interaction of wave-numbers of the same order of magnitude, since u^2 is the order of magnitude of Reynolds stresses produced by the energy-containing range of wave-numbers and u/l is the order of magnitude of the rate of shearing produced by that same range. Equivalently,

$$\frac{\mathrm{d}u^2}{\mathrm{d}t} = -Aul\frac{u^2}{l^2}$$

suggests that an eddy viscosity of order ul is acting on the shear of order u/l to produce a 'dissipation' of energy from the energy-containing eddies to smaller eddies. It seems that the energy-containing eddies determine the rate of energy transfer by their mutual inertial interaction, and the larger wave-numbers adjust themselves, according to the Reynolds number, in order to convert this energy into heat at the required rate. The part played by viscous forces in the motion at these high Reynolds numbers is entirely secondary, and may be ignored for all except the high wave-numbers at which the dissipation occurs.

It can now be argued that if the Fourier coefficients for high wave-numbers are statistically independent of those for the energy-containing range, the former *must* describe a motion which is in statistical equilibrium since no time-dependence can be imposed on them. Hence the hypothesis of this section can be regarded as including the hypothesis of statistical equilibrium put forward in §6.1. However, the latter is very plausible on its own merits and makes a natural starting-point.

6.4. The universal equilibrium

The considerations in the last three sections have prepared the ground for a statement of the theory of the universal equilibrium—termed the theory of local similarity by its originator, A. N. Kolmogoroff (1941 *a*, *c*), and also referred to as the theory of similarity of the small eddies. We have seen that when the Reynolds number is sufficiently large there is likely to exist a range of (high) wave-numbers which is responsible for most of the viscous dissipation and for which the Fourier coefficients are statistically steady, isotropic and independent of the Fourier coefficients of the energy-containing range of wave-numbers. We shall call this the equilibrium range of wave-numbers. On what does the motion associated with this equilibrium range depend? It is internally self-adjusting, through the operation of inertia forces, and must depend only on the parameters which describe external effects. These external effects are just two in number: the removal of energy by viscous dissipation over the whole, but chiefly at the upper end, of the equilibrium range and the insertion of energy by inertial transfer at the lower end of the range. The removal and insertion of energy proceed at the same rate, viz.

$$\epsilon = -\frac{3}{2}\frac{du^2}{dt} = 2\nu\int_0^\infty \kappa^2 E(\kappa, t)\,d\kappa,$$

which must be one of the external parameters. The removal of energy by dissipation is distributed over the equilibrium range in a way which depends on the viscosity ν (as well as on the spectrum function $E(\kappa, t)$, of course), which is also an external parameter. No other parameters are needed to specify the equilibrium range in view of the hypothesis of statistical independence put forward in the previous section. Hence we have the hypothesis of universal equilibrium:

'The motion associated with the equilibrium range of wave-numbers is uniquely determined statistically by the parameters ϵ and ν.'

The reason for referring to the equilibrium of the large wave-numbers as *universal* is now clear. On dimensional grounds the effect of variation of the parameters ϵ and ν can only be to change the effective length and time scales of the motion. It is usually more convenient to regard the length and velocity scales as variable, so that we define the basic length and velocity parameters

$$\eta=\left(\frac{\nu^3}{\epsilon}\right)^{\frac{1}{4}}, \quad v=(\nu\epsilon)^{\frac{1}{4}}. \tag{6.4.1}$$

When all lengths are referred to η as unit and all velocities to v, the motion associated with the equilibrium range of wave-numbers thus has a universal statistical form.

In particular, the statistical quantities determined by the equilibrium range are independent of the properties of the large-scale components of the turbulence, and do not require the turbulence to be accurately homogeneous. If the departure from homogeneity has the same length scale as the energy-containing eddies of the turbulence, as is usually the case inasmuch as the geometry of the boundaries of the fluid determines both scales, the whole of the argument of this chapter applies equally well. The motion associated with the equilibrium range is still approximately homogeneous owing to the smaller size of the eddies concerned, and the only difference in such a case is that the value of ϵ to be used in (6.4.1) is local in space as well as in time. The hypotheses of this and the previous section thus have an importance which goes beyond our special case of homogeneous turbulence. If the hypotheses are valid, all turbulent motions—decaying homogeneous turbulence, flow in a pipe under pressure, flow in a boundary layer, turbulent wakes, flow of a fluid with density stratification, etc.—are such that at sufficiently large Reynolds number the motions associated with the small eddies have a common statistical form. However, sufficient experimental evidence relating to these different types of turbulent motion is not yet available, so that this aspect of the theory must be regarded cautiously.

In analytical terms, the hypothesis of this section is that at sufficiently high Reynolds number the joint probability distribution of the values of the Fourier coefficient $d\mathbf{Z}(\boldsymbol{\kappa}, t)$ at any n values of $\boldsymbol{\kappa}$ is independent of t in form, is of isotropic form, and is universal when v and η are used as velocity and length units, provided each of the n values of κ is sufficiently large. The necessary restrictions on the Reynolds number and on κ can be made a little more precise.

We require the Reynolds number to be large enough for a statistically independent equilibrium range to exist, and it was seen in the previous section that a necessary condition for this is $\kappa_0 \ll \kappa_d$ (see (6.3.3)). The wave-number κ_d which marks the location of strong viscous forces must be of the order of magnitude of $1/\eta$ (for, if the theory is correct, there is no other length relevant to the equilibrium with which η can be identified;† in the sequel we shall regard κ_d as being defined as exactly equal to $1/\eta$). Hence the Reynolds number must be large enough to satisfy

$$\kappa_0 \ll \frac{1}{\eta} = \left(\frac{\epsilon}{\nu^3}\right)^{\frac{1}{4}}.$$

Replacing κ_0 by $1/l$ and making use of (6.1.1), which was introduced as an empirical relation and which is now an integral part of the theory, we require

$$\frac{1}{l} \ll \left(\frac{u^3}{l\nu^3}\right)^{\frac{1}{4}}, \quad \text{i.e.} \quad \left(\frac{ul}{\nu}\right)^{\frac{3}{4}} \gg 1. \qquad (6.4.2)$$

This necessary condition is not far from being simply that the Reynolds number of the energy-containing eddies should be large compared with unity—in other words, that the motion associated with the energy-containing range of wave-numbers should be entirely dominated by inertia forces, as indeed we postulated at the beginning.

Similarly, a necessary restriction on κ for the above hypothesis to be valid is $\kappa \gg \kappa_0$, since the motion associated with wave-number κ would not otherwise be statistically independent of the energy-containing range. Our two conditions are therefore

$$\left(\frac{ul}{\nu}\right)^{\frac{3}{4}} \gg 1, \quad \kappa \gg \kappa_0 = \frac{1}{l}. \qquad (6.4.3)$$

The numerical interpretation of the symbol $\gg$ must remain to be determined experimentally, since our arguments have shown only the *necessity* for certain ratios to be large; the determination of *sufficient* conditions for statistical equilibrium, or for statistical independence of the Fourier coefficients at two different wave-

† Moreover, the Reynolds number $v\eta/\nu$ is identically unity, showing that inertia and viscous forces are comparable for the motion for which v and η are a representative velocity and length.

numbers, is a much more difficult problem which is unlikely to be solved by the kind of argument used in this chapter.

With the aid of the universal equilibrium hypothesis we can proceed to make predictions about any statistical quantities which depend only on the values of the Fourier coefficients $d\mathbf{Z}(\boldsymbol{\kappa}, t)$ in the equilibrium range of wave-numbers. The energy spectrum function $E(\kappa, t)$ depends only on Fourier coefficients for wave-numbers with magnitude κ, so that provided the conditions (6.4.3) are satisfied we have the prediction

$$E(\kappa, t) = v^2\eta E_e(\eta\kappa), \tag{6.4.4}$$

where E_e is a dimensionless function of universal form. We note that (6.4.4) is consistent with the original assumption that the dissipation occurs wholly in the equilibrium range, since

$$2\nu \int_0^\infty \kappa^2 E(\kappa, t)\, d\kappa = \epsilon$$

is satisfied identically provided the universal function E_e satisfies

$$\int_0^\infty y^2 E_e(y)\, dy = \tfrac{1}{2}. \tag{6.4.5}$$

Direct measurements of $E(\kappa, t)$ at large values of κ for turbulence at very large Reynolds numbers would provide the best evidence for the validity of (6.4.4) and of the equilibrium hypothesis. However, Reynolds numbers (of turbulence generated by grids) such that the energy-containing range and the dissipation range are widely separated are beyond the limit attainable in most of the wind tunnels used in laboratories, and measurements under the required conditions have not yet been made. An indication of a trend towards the prediction (6.4.4) as the Reynolds number is increased is provided by some published measurements (Batchelor and Townsend, 1949) of the dimensionless ratio

$$\frac{\overline{\left(\frac{\partial u_1}{\partial x_1}\right)^2}\;\overline{\left(\frac{\partial^3 u_1}{\partial x_1^3}\right)^2}}{\left[\overline{\left(\frac{\partial^2 u_1}{\partial x_1^2}\right)^2}\right]^2} = \frac{35}{27}\,\frac{\int_0^\infty \kappa^6 E(\kappa, t)\, d\kappa \int_0^\infty \kappa^2 E(\kappa, t)\, d\kappa}{\left[\int_0^\infty \kappa^4 E(\kappa, t)\, d\kappa\right]^2} \tag{6.4.6}$$

under different conditions. When the Reynolds number is large enough for the equilibrium range to exist, this ratio is determined by the function E_e only and should therefore be an absolute constant.

Six of the points in fig. 6.3† represent measurements for turbulence produced by a grid, each of which is approximately independent of

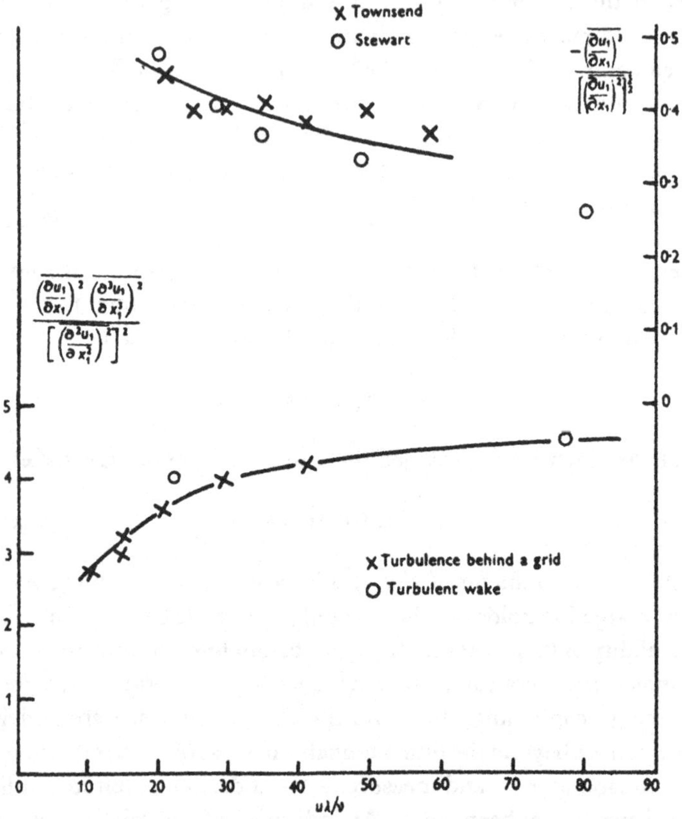

Fig. 6.3. Variation of dimensionless ratios with Reynolds number (after Batchelor and Townsend, 1947, 1949; and Stewart, 1951).

decay of the turbulence; the other two points represent measurements in the central plane of the turbulent wake behind a cylinder. The measurements are consistent with the validity of the theoretical

† The most appropriate abscissa for fig. 6.3 would be ul/ν, but as measurements of a length characteristic of the energy-containing eddies were not made, $u\lambda/\nu$ has been used. The relation between these Reynolds numbers for isotropic turbulence is obtained from (5.5.6) and (6.1.1) as

$$\frac{ul}{\nu} = \frac{A}{10}\left(\frac{u\lambda}{\nu}\right)^2,$$

and a roughly similar relation will apply in the case of the turbulent wake.

prediction, but higher Reynolds numbers are needed for a more decisive test. The values of ul/ν for the points plotted in fig. 6.3 range from about 10·4 to 540.

In the case of mean values which are defined as integrals (with respect to $\boldsymbol{\kappa}$) of the Fourier coefficients $\mathrm{d}\mathbf{Z}(\boldsymbol{\kappa}, t)$ (in contradistinction to $E(\kappa, t)$ which depends on a local value of $\mathrm{d}\mathbf{Z}(\boldsymbol{\kappa}, t)$ only), it is not easy to decide whether or not these mean values are determined by the equilibrium range of wave-numbers. A useful criterion in such cases is provided by an attempt to express the mean value wholly in terms of the *difference* of the velocities at two neighbouring points. If the points are $\mathbf{x}$ and $\mathbf{x}+\mathbf{r}$, we have

$$\mathbf{u}(\mathbf{x}+\mathbf{r}, t) - \mathbf{u}(\mathbf{x}, t) = \int e^{i\boldsymbol{\kappa}.\mathbf{x}}(e^{i\boldsymbol{\kappa}.\mathbf{r}} - 1)\,\mathrm{d}\mathbf{Z}(\boldsymbol{\kappa}, t); \qquad (6.4.7)$$

consequently a mean value formed from $\mathbf{u}(\mathbf{x}+\mathbf{r}, t) - \mathbf{u}(\mathbf{x}, t)$ weights the Fourier coefficients with the factor $e^{i\boldsymbol{\kappa}.\mathbf{r}} - 1$ by comparison with the corresponding mean value formed from $\mathbf{u}(\mathbf{x}, t)$. When $|\boldsymbol{\kappa}.\mathbf{r}| \ll 1$ the weighting factor is approximately linear in κ. It is exactly linear in the integral giving the dissipation (which depends on the velocity derivatives), so that if r is such that $r\kappa_0 \ll 1$, the Fourier coefficients for the energy-containing range of wave-numbers are suppressed as strongly as in the expression for the dissipation. This is clearly a sufficient condition for the mean value to be determined by the equilibrium range (when the latter exists). Since the weighting factor $(e^{i\boldsymbol{\kappa}.\mathbf{r}} - 1)$ is not small when κ is of order $1/r$, the above two conditions for the equilibrium range, viz. $\kappa \gg \kappa_0$ and $r \ll 1/\kappa_0$, are consistent provided we give the same numerical meaning to the sign $\gg$ in the two cases.

An example of a measurable mean value which depends on velocity derivatives only and which therefore comes within the scope of the equilibrium theory is $\overline{(\partial u_1/\partial x_1)^3} = u^3 k_0'''$. At sufficiently large Reynolds numbers the dimensionless ratio

$$\overline{\left(\frac{\partial u_1}{\partial x_1}\right)^3} \Big/ \left[\overline{\left(\frac{\partial u_1}{\partial x_1}\right)^2}\right]^{\frac{3}{2}}$$

should thus be an absolute constant. Measurements for turbulence produced by a grid (Batchelor and Townsend, 1947; Stewart, 1951) are shown in fig. 6.3, and are again consistent with, but do not go far enough to confirm, the theoretical prediction.

There is an obvious dualism of the Fourier coefficients and the velocity differences, and the universal similarity hypothesis may be formulated to refer to mean values formed either from the Fourier coefficients $\mathrm{dZ}(\boldsymbol{\kappa}, t)$ for $\kappa \gg \kappa_0$, or from velocity differences

$$\mathbf{u}(\mathbf{x}+\mathbf{r}, t)-\mathbf{u}(\mathbf{x}, t) \quad \text{for} \quad r \ll \frac{1}{\kappa_0}.$$

The results from one form of the hypothesis may usually be recovered from those from the other form with the aid of a Fourier transform relation. For instance, from (3.4.14) and (3.4.15),

$$\overline{|\mathbf{u}(\mathbf{x}+\mathbf{r}, t)-\mathbf{u}(\mathbf{x}, t)|^2} = 6u^2 - 2R_{ii}(\mathbf{r}, t)$$
$$= 4\int_0^\infty E(\kappa, t)\left(1-\frac{\sin \kappa r}{\kappa r}\right) d\kappa, \qquad (6.4.8)$$

and if $\kappa_0 r \ll 1$ this integral depends on the equilibrium range of κ only (with an absolute error of the same order as that made in the assumption that the dissipation integral does not depend on the energy-containing range), in which case we have from (6.4.4)

$$\overline{|\mathbf{u}(\mathbf{x}+\mathbf{r}, t)-\mathbf{u}(\mathbf{x}, t)|^2} = 4v^2\int_0^\infty E_e(y)\left(1-\frac{\sin yr/\eta}{yr/\eta}\right) dy \qquad (6.4.9)$$
$$= v^2 \times \text{universal function of } r/\eta.$$

This is exactly the prediction that would have been made had we postulated that the joint probability distribution of the values of $\mathbf{u}(\mathbf{x}+\mathbf{r}, t)-\mathbf{u}(\mathbf{x}, t)$ at any n values of $\mathbf{r}$ is universal when v and η are used as units, provided each of the values of $\mathbf{r}$ satisfies $r \ll l$, as was done originally by Kolmogoroff (1941 *a*).† Formulation of the universal similarity hypothesis in terms of Fourier coefficients seems to have the advantage that we are aided by the interpretation of

† Kolmogoroff, in fact, made the more general postulate (which is consistent with the basic physical ideas already put forward) that the joint-probability distribution of the values of $\mathbf{u}(\mathbf{x}+\mathbf{r}+\mathbf{s}, t+\tau)-\mathbf{u}(\mathbf{x}, t)$ at any n values of $(\mathbf{r}, \tau)$, each of which satisfies $r \ll l$ and $\tau \ll l/u$, is universal when v and η are used as units, provided the Reynolds number is large enough. The (random) vector $\mathbf{s}$ is equal to $\tau\mathbf{u}(\mathbf{x}, t)$ and is introduced in order to remove the effect, on the velocity difference, of translation of the small-eddy flow pattern by energy-containing eddies during the interval τ; the velocity difference is thereby converted to a kind of Lagrangian, or particle, quantity. To make an equivalent statement about Fourier coefficients it would first be necessary to make a Fourier analysis of $\mathrm{dZ}(\boldsymbol{\kappa}, t)$ with respect to t (which is carried out in the manner already described, since $\mathrm{dZ}(\boldsymbol{\kappa}, t)$ is approximately a stationary random function of t when $\kappa \gg \kappa_0$). However, time interval mean values play no essential part in the existing analysis of turbulence—except in problems of diffusion, which are outside the scope of this work—and will not be considered here.

Fourier components as degrees of freedom of the motion. On the other hand, the velocity differences have a more immediate physical meaning, and are usually more directly related to the observed quantities.

A typical application of the universal equilibrium theory has been made by Kolmogoroff (1949). With arguments like those which lead to (6.4.4), he obtains information about the size of drops of foreign liquid which exist in a fluid in turbulent motion under the opposing actions of surface tension and the shearing produced by the turbulence.

6.5. The inertial subrange

It has been postulated that the universal equilibrium will exist when the Reynolds number is large enough for the energy-containing and dissipation ranges of wave-numbers to be widely separated; presumably a sufficient condition is that κ_d/κ_0 should exceed some critical (large) value. Since all the dissipation necessarily occurs in the equilibrium range, this is equivalent to assuming that the equilibrium range (when it exists) always *begins* at a wave-number κ such that κ/κ_0 has this critical value and extends to $\kappa=\infty$, irrespective of where the dissipation occurs in the equilibrium range. Consequently, when the Reynolds number is large enough, there may exist a considerable subrange, within and at the lower end of the equilibrium range, in which negligible dissipation occurs. Within this inertial subrange, the transfer of energy by inertia forces is the dominant process. A necessary condition for the existence of the inertial subrange is evidently that the Reynolds number should be high enough for it to be possible to find wave-numbers κ such that

$$\kappa_0 \ll \kappa \ll \kappa_d,$$

i.e.

$$\frac{1}{l} \ll \kappa \ll \frac{1}{\eta}. \tag{6.5.1}$$

We have seen (see (6.4.2)) that l/η is of order $(ul/\nu)^{\frac{3}{4}}$, so that the condition (6.5.1) is equivalent to

$$\left(\frac{ul}{\nu}\right)^{\frac{3}{4}} \gg 1, \tag{6.5.2}$$

and when it is satisfied wave-numbers of the order of $\left(\frac{1}{l\eta}\right)^{\frac{1}{2}}$, that is, of order $\frac{1}{l}\left(\frac{ul}{\nu}\right)^{\frac{3}{4}}$, will lie within the inertial subrange. The existence of the inertial subrange requires the Reynolds number to be of the order of the square of that required for the existence of the equilibrium range, and is clearly unlikely to be realized in laboratory wind tunnels.

When the condition (6.5.1) is satisfied, the motion associated with the inertial subrange is statistically independent, according to the hypothesis of § 6.3, of both the energy-containing eddies and the eddies responsible for the dissipation. The motion associated with the inertial subrange is therefore determined uniquely when the rate at which energy is entering it at the low wave-number end and leaving it at the high wave-number end, viz. ϵ, is specified. This is a very restrictive consequence of our hypotheses, and enables us to make very specific predictions about mean values determined by the inertial range of wave-numbers. The universal functions found for the equilibrium range generally are required to take a form, in the inertial subrange, such that the parameter ν disappears from the expression for the mean value.

When the Reynolds number is high enough for the equilibrium range to exist, we found the form $v^2\eta E_e(\eta\kappa)$ (see (6.4.4)) for the energy spectrum function $E(\kappa, t)$ in the range $\kappa \gg \kappa_0$. If it is also true that values of κ such that $\kappa_0 \ll \kappa \ll \kappa_d$ can be found, then for these values of κ the function E_e must take a form such that $v^2\eta E_e(\eta\kappa)$ is independent of ν. With the aid of (6.4.1) we thus find

$$E(\kappa, t) = \alpha v^2 \eta(\eta\kappa)^{-\frac{5}{3}} = \alpha\epsilon^{\frac{2}{3}}\kappa^{-\frac{5}{3}} \tag{6.5.3}$$

for $\kappa_0 \ll \kappa \ll \kappa_d$, where α is an absolute constant.

If the universal equilibrium hypothesis is formulated in terms of the probability distribution of velocity differences rather than Fourier coefficients, there is likewise an inertial subrange if the velocities are taken at points separated by an interval r such that

$$l \gg r \gg \eta. \tag{6.5.4}$$

For such values of r, the mean values formed from velocity differences must depend on ϵ alone; for instance, we find that

$$\overline{|\mathbf{u}(\mathbf{x}+\mathbf{r}, t) - \mathbf{u}(\mathbf{x}, t)|^2} = \beta v^2 \left(\frac{r}{\eta}\right)^{\frac{2}{3}} = \beta(\epsilon r)^{\frac{2}{3}} \tag{6.5.5}$$

for $l \gg r \gg \eta$, where β is an absolute constant.

The predictions (6.5.3) and (6.5.5) are readily found to be consistent from a consideration of the exact relation

$$\overline{|\mathbf{u}(\mathbf{x}+\mathbf{r},t)-\mathbf{u}(\mathbf{x},t)|^2}=4\int_0^\infty E(\kappa,t)\left(1-\frac{\sin\kappa r}{\kappa r}\right)d\kappa.$$

Provided r satisfies (6.5.4), the integrand is suppressed by the small value of the factor $\left(1-\frac{\sin\kappa r}{\kappa r}\right)$ when κ is of order $1/l$, and is suppressed by the small value of $E(\kappa,t)$ when κ is of order $1/\eta$ (since $E(\kappa,t)$ falls off sharply at wave-numbers where viscous forces are dominant). The value of the integral is therefore dominated by the behaviour of the integrand for values of κ such that $\kappa_0 \ll \kappa \ll \kappa_d$, and hence

$$\overline{|\mathbf{u}(\mathbf{x}+\mathbf{r},t)-\mathbf{u}(\mathbf{x},t)|^2}\approx 4\int_0^\infty \alpha\epsilon^{\frac{2}{3}}\kappa^{-\frac{5}{3}}\left(1-\frac{\sin\kappa r}{\kappa r}\right)d\kappa$$

$$=\tfrac{9}{5}\Gamma(\tfrac{1}{3})\,\alpha(\epsilon r)^{\frac{2}{3}}. \qquad (6.5.6)$$

The relation between the absolute constants in (6.5.3) and (6.5.5) is evidently

$$\beta=\tfrac{9}{5}\Gamma(\tfrac{1}{3})\alpha=4{\cdot}82\alpha. \qquad (6.5.7)$$

In the limit of infinite Reynolds number, in which case there are some values of κ and r such that κ/κ_0, κ_d/κ, r/η and l/r are all infinitely large, the relations (6.5.3), (6.5.5) and (6.5.6) become exact.

As already mentioned, it is not likely that the conditions necessary for these results to be valid will be realized in wind tunnels in a laboratory. A grid with bars of diameter $M/5{\cdot}3$ spaced at distance M apart (in two perpendicular directions), placed in a uniform stream of air with speed U, will generate turbulence whose Reynolds number (based on the longitudinal integral scale) is

$$\frac{uL_p}{\nu}\approx\frac{1}{10}\left(\frac{u\lambda}{\nu}\right)^2\approx\frac{1}{134}\frac{UM}{\nu}$$

(Batchelor and Townsend, 1948*a*). Taking $\nu=0{\cdot}15\,\text{cm}^2\,\text{sec}^{-1}$, $M=5$ cm, $U=2000$ cm sec^{-1} (i.e. $UM/\nu=6{\cdot}7\times10^4$), which are values of M and U probably as large as can conveniently be used in a laboratory wind tunnel, we find $(uL_p/\nu)^{\frac{3}{4}}=105$, which might perhaps be large enough† for the equilibrium range to exist, but is not large enough to satisfy the necessary condition (6.5.2) for the existence of the inertial subrange. Since, in addition, the functional forms (6.5.3) and (6.5.5) will be detectable empirically only if they are valid over an appreciable range of values of κ, it is clear‡ that we cannot look to laboratory experiments for confirmation of the predictions about the inertial subrange.

If the universal equilibrium theory as a whole is valid—and the evidence available at the present time is favourable—the predictions about the inertial subrange should find many important applications in the fields of meteorology, oceanography and astrophysics. Turbulent motions play a prominent part in these fields and the corresponding Reynolds numbers are very large. In the case of atmospheric turbulence, casual observation by eye suggests figures in the neighbourhood of $l = 100$ metres, $u = 0{\cdot}5$ metre sec^{-1}, at heights up to about 10 km, for at least some synoptic conditions (these figures are also consistent with (6.1.1) and Brunt's estimate§ that ϵ is of the order of 5 cm^2 sec^{-3}). The corresponding value of $(ul/\nu)^{\frac{3}{4}}$ is $7{\cdot}8 \times 10^4$, which is probably large enough to satisfy the condition for the existence of the inertial subrange, the values of r defining this range being given by

$$100 \text{ metres (or better estimate of } l) \gg r \gg (\nu^3/\epsilon)^{\frac{1}{4}} = 0{\cdot}2 \text{ cm.}$$

Thus predictions about the inertial subrange are not only very valuable on account of their specificness, but also because they concern a range of eddy sizes which is likely to be relevant to many of the measurable effects of turbulence. The same remarks apply to turbulent motions encountered in astrophysics, as, for instance, in the atmosphere of the sun. For all large-scale turbulent motions, the theory described in this section can make very specific predictions about any mean value which depends on eddies which are sufficiently small compared with the energy-containing eddies. As

† Apparently not, according to some recent calculations by R. W. Stewart (1951), for a slightly smaller Reynolds number; Stewart showed, from measurements of the triple-velocity correlation $u^3k(r)$, that the rate at which all Fourier components with wave-number magnitudes greater than κ' were receiving energy by inertial transfer was only about one-half the rate of loss by viscous dissipation, for values of κ' in the dissipation range.

‡ Despite many attempts in the past, including some by the author.

§ D. Brunt, *Physical and Dynamical Meteorology*, Cambridge University Press, 1944, p. 286.

already shown, the only analysis required is a dimensional argument. Applications of the theory of the inertial subrange which have already been made (not all of them being uncontentious) concern the functional form of the covariance of the fluctuating temperatures at two points (Obukhoff, 1949*b*; Inoue, 1950*b*; Corrsin, 1951*b*), the functional form of the pressure covariance (Inoue, 1950*a*), the relative diffusion of two particles of fluid (Batchelor, 1950*a*, 1952*a*), the functional form of the Lagrangian velocity covariance for a single fluid particle (Inoue, 1950*c*, 1951*a*), and the rate at which lines moving with the fluid are extended in length (Batchelor, 1950*b*, 1952*c*), none of which will be described here.

6.6. The energy spectrum in the equilibrium range

The theoretical determination of the energy spectrum function $E(\kappa, t)$ over the whole of the equilibrium range of wave-numbers is of considerable importance, particularly for turbulent motions set up in the laboratory for which the inertial subrange is of negligible extent. Conditions in the equilibrium range are comparatively simple and offer us the best prospect of being able to take account of the energy transfer quantitatively and to determine $E(\kappa, t)$. If the spectrum shape can be determined, then, according to the foregoing theory, it will apply to all kinds of turbulent motion, whatever their large-scale properties, the only condition being that the Reynolds number should be sufficiently large.

Several attempts to calculate the spectrum on the basis of intuitive hypotheses about the transfer of energy have been made. There are no measurements which support any of these theoretical spectrum functions, and the hypotheses underlying them must be regarded as very speculative, but they have some intrinsic interest and will be described briefly in this section. It should be noted that none of the hypotheses to be described are *consequences* of the universal equilibrium theory, and the failure of the former would in no way compromise the position of the equilibrium theory. The only reason for considering these hypotheses about the energy transfer in the present chapter is that they are made *possible* by the simple results derived from the equilibrium theory.

Since the turbulence in the equilibrium range of κ is isotropic, the dynamical equation for the energy spectrum tensor for large values of κ is equivalent to the single scalar equation

$$\frac{\partial E(\kappa, t)}{\partial t} = T(\kappa, t) - 2\nu\kappa^2 E(\kappa, t); \tag{6.6.1}$$

$T(\kappa, t)$ is the contribution from inertial transter of energy and is given by (see (5.5.19))

$$T(\kappa, t) = 4\pi\kappa^2 \int Q(\boldsymbol{\kappa}, \boldsymbol{\kappa}', t)\, d\boldsymbol{\kappa}', \tag{6.6.2}$$

where Q is the exchange function defined in terms of the mean product of three Fourier coefficients by (5.2.7). We shall find it easier to think about the equation (6.6.1) in the form (5.5.17), viz.

$$\frac{\partial \int_0^\kappa E(\kappa'', t)\, d\kappa''}{\partial t} = -S(\kappa, t) - 2\nu \int_0^\kappa \kappa''^2 E(\kappa'', t)\, d\kappa'', \tag{6.6.3}$$

where
$$S(\kappa, t) = \int_{\kappa'=\kappa}^{\infty} \int_{\kappa''=0}^{\kappa} P(\kappa', \kappa'', t)\, d\kappa'\, d\kappa'', \tag{6.6.4}$$

and P is related to Q by (5.5.18); $S(\kappa, t)$ is the net rate of transfer of energy from wave-number magnitudes less than κ to those greater than κ at time t.

According to the universal equilibrium theory we can ignore the dependence on time of mean values determined by the equilibrium range of wave-numbers, so that when $\kappa \gg \kappa_0$, and for $(ul/\nu)^{\frac{1}{2}} \gg 1$, (6.6.1) reduces to

$$T(\kappa) = 2\nu\kappa^2 E(\kappa). \tag{6.6.5}$$

Equivalently, we can replace $\frac{\partial}{\partial t}\int_0^\kappa E(\kappa'', t)\, d\kappa''$ by the total rate of decay of energy, $-\epsilon$, since a negligible amount of energy is contained in wave-number magnitudes greater than κ. Hence (6.6.3) becomes

$$2\nu \int_\kappa^\infty \kappa''^2 E(\kappa'')\, d\kappa'' = S(\kappa) = \int_{\kappa'=\kappa}^{\infty} \int_{\kappa''=0}^{\kappa} P(\kappa', \kappa'')\, d\kappa'\, d\kappa''. \tag{6.6.6}$$

If either (6.6.5) or (6.6.6) is to be solved for $E(\kappa)$, we must make a plausible guess at the form of one of the transfer functions $T(\kappa)$ and $S(\kappa)$. All the exact information which we have available for guidance is contained in the above equations, together with (5.5.18) and (5.2.7). It will be sufficient to concentrate on (6.6.6), and all that we know about the exchange function $P(\kappa', \kappa'')$ is that it is antisymmetrical in κ' and κ'', that (according to the independence hypothesis of § 6.3) it falls to zero when $\kappa' \ll \kappa''$ or $\kappa' \gg \kappa''$, and that it is related to the Fourier coefficients through (5.5.18) and (5.2.7).

From a mathematical point of view we should like to be able to express $P(\kappa', \kappa'')$ in terms of the spectrum function E, since equation (6.6.6) would then suffice to determine $E(\kappa)$. In general $P(\kappa', \kappa'')$ will not depend *only* on the spectrum function, but within the equilibrium range $P(\kappa', \kappa'')$ will be determined by the function E in the formal sense that the whole probability distribution has a universal form. Our task is to postulate, intuitively, a suitable approximate relation between $P(\kappa', \kappa'')$ and the function E. The relation between $P(\kappa', \kappa'')$ and the Fourier coefficients $\mathrm{d}\mathbf{Z}$, given by (5.5.18) and (5.2.7), suggests that $P(\kappa', \kappa'')$ will depend on $E(\kappa')$, $E(\kappa'')$ and $E(|\boldsymbol{\kappa}' - \boldsymbol{\kappa}''|)$, and possibly on κ' and κ''.

Perhaps the closest approach to these requirements, while retaining enough simplicity to make the equation (6.6.6) soluble, is made by the expression put forward by T. von Kármán (1948*a*), viz.

$$P(\kappa', \kappa'') = 2\gamma \kappa'^{m} \kappa''^{\frac{1}{2}-m} [E(\kappa')]^{n} [E(\kappa'')]^{\frac{3}{2}-n} \quad (\kappa' > \kappa''), \qquad (6.6.7)$$

which is dimensionally correct for arbitrary values of m and n; the number γ is an absolute constant. It is a serious criticism of this expression that it does not contain $E(|\boldsymbol{\kappa}' - \boldsymbol{\kappa}''|)$, which would seem, according to (5.2.7), to have as strong a claim for inclusion as either $E(\kappa')$ or $E(\kappa'')$. A related defect is that the expression (6.6.7) changes sign discontinuously at $\kappa' = \kappa''$, whereas (5.2.7) probably does so smoothly; however, this is more a defect of the degree of approximation and is less fundamental than that mentioned first. The advantage of the form (6.6.7) lies in the separation of the variables κ' and κ'', which allows (6.6.6) to be reduced to

$$2\nu \int_{\kappa}^{\infty} \kappa''^{2} E(\kappa'') \,\mathrm{d}\kappa'' = 2\gamma \int_{\kappa}^{\infty} \kappa'^{m} [E(\kappa')]^{n} \,\mathrm{d}\kappa' \int_{0}^{\kappa} \kappa''^{\frac{1}{2}-m} [E(\kappa'')]^{\frac{3}{2}-n} \,\mathrm{d}\kappa''. \qquad (6.6.8)$$

We note that when κ is in the inertial subrange, the left-hand side reduces to the constant value ϵ, and provided m and n have values such that the integrals on the right-hand side are convergent at the fixed terminals, a solution is

$$E(\kappa)=\left(\frac{\epsilon}{2\gamma}\right)^{\frac{2}{3}}\left(m-\tfrac{5}{3}n+1\right)^{\frac{4}{3}}\kappa^{-\frac{5}{3}} \qquad (6.6.9)$$

(the restriction on m and n being that $m-\frac{5}{3}n+1<0$). This result is in agreement with that already found in (6.5.3), but it is in consequence of dimensional arguments alone and the agreement cannot be used as a test of the validity of (6.6.7). Provided the integral $\int_0^{\kappa}\kappa''^{\frac{1}{2}-m}[E(\kappa'')]^{\frac{3}{2}-n}\,d\kappa''$ converges as $\kappa\to\infty$, the solution for very large values of κ is of the form

$$E(\kappa)\sim\kappa^{-(m-2)/(n-1)} \qquad (6.6.10)$$

$\left(\text{the condition for convergence then being that } -\frac{m-2}{n-1}<1\right)$. This asymptotic variation of $E(\kappa)$ according to a power law implies that some of the higher order derivatives of the velocity correlation function do not exist at $r=0$, and since

$$\overline{\frac{\partial^p u_1}{\partial x_1^p}\frac{\partial^q u_1}{\partial x_1^q}}=(-1)^p\left[\frac{\partial^{2s}u^2 f(r)}{\partial r^{2s}}\right]_{r=0} \qquad (p+q=2s),$$

this in turn implies that some of the higher order derivatives of $\mathbf{u}(\mathbf{x})$ do not exist in general. A proof that solutions of the Navier-Stokes equation possess derivatives of all orders has not been given, but one suspects that there is a real inconsistency.

An exchange function which is included in the formula (6.6.7) and which can be given some support by intuitive arguments has been put forward by C. F. von Weizsäcker (1948) and formulated analytically by W. Heisenberg (1948*a*). This exchange function corresponds to the case $m=-\frac{3}{2}$, $n=\frac{1}{2}$, and the corresponding transfer function is

$$S(\kappa)=\gamma\int_{\kappa}^{\infty}\kappa'^{-\frac{3}{2}}[E(\kappa')]^{\frac{1}{2}}\,d\kappa'\int_0^{\kappa}2\kappa''^2E(\kappa'')\,d\kappa''. \qquad (6.6.11)$$

The idea underlying this postulate is that the process of transfer of energy from large to small eddies is qualitatively similar to the conversion of mechanical energy in a fluid into thermal energy through the agency of molecular motion. This latter rate of transfer of energy is equal to the molecular viscosity multiplied by the mean-square vorticity of the mass motion of the fluid. Hence we might expect the rate of transfer from large to small eddies to be given by the product of an effective viscosity produced by the motion of the small eddies and the mean-square vorticity associated with the large eddies, giving

$$S(\kappa) = \mathrm{N}(\kappa)\int_0^\kappa 2\kappa''^2 E(\kappa'')\,\mathrm{d}\kappa''. \tag{6.6.12}$$

If each small element of the range of wave-numbers from $\kappa' = \kappa$ to $\kappa' = \infty$ is to make a separate and similar contribution to the effective viscosity $\mathrm{N}(\kappa)$ which depends on the energy density $E(\kappa')$ and the wave-number κ' only, dimensional considerations show that

$$\mathrm{N}(\kappa) = \gamma\int_\kappa^\infty \kappa'^{-\frac{3}{2}}[E(\kappa')]^{\frac{1}{2}}\,\mathrm{d}\kappa', \tag{6.6.13}$$

which leads to the expression (6.6.11). The notion that the small eddies act as an effective viscosity is plausible enough in general, but it does not seem a suitable description of the mutual action of eddies whose sizes are of the same order of magnitude. We have seen earlier in this chapter that the transfer of energy across the wave-number κ occurs principally—at all events, when inertia forces are dominant—as a result of interaction of Fourier components with wave-numbers of the same order as κ, so that the accuracy of Heisenberg's expression for the transfer cannot immediately be granted.

With the aid of (6.6.11), the equation (6.6.6) can be written as

$$\epsilon = \left[\nu + \gamma\int_\kappa^\infty \kappa'^{-\frac{3}{2}}[E(\kappa')]^{\frac{1}{2}}\,\mathrm{d}\kappa'\right]\int_0^\kappa 2\kappa''^2 E(\kappa'')\,\mathrm{d}\kappa'', \tag{6.6.14}$$

which shows the roles of the molecular viscosity and the viscosity arising from the motion of the small eddies. The solution (pointed out first by J. Bass, 1949) can be found readily by the device of using

$$H(\kappa) = \int_0^\kappa \kappa''^2 E(\kappa'')\,\mathrm{d}\kappa''$$

as dependent variable, whence (6.6.14) becomes

$$\epsilon = 2\left[\nu + \gamma\int_\kappa^\infty \kappa'^{-\frac{5}{2}}\left[\frac{\mathrm{d}H(\kappa')}{\mathrm{d}\kappa'}\right]^{\frac{1}{2}}\mathrm{d}\kappa'\right]H(\kappa).$$

On dividing by $H(\kappa)$ and differentiating, we find

$$\frac{dH(\kappa)}{d\kappa} = \frac{4\gamma^2}{\epsilon^2}\frac{[H(\kappa)]^4}{\kappa^5},$$

of which the solution is

$$[H(\kappa)]^{-3} = \frac{3\gamma^2}{\epsilon^2}\kappa^{-4} + \left(\frac{\epsilon}{2\nu}\right)^{-3},$$

since $H(\kappa) \to \epsilon/2\nu$ as $\kappa \to \infty$. Hence

$$E(\kappa) = \frac{1}{\kappa^2}\frac{dH(\kappa)}{d\kappa} = \left(\frac{8\epsilon}{9\gamma}\right)^{\frac{2}{3}} \kappa^{-\frac{5}{3}} \left[1 + \frac{8\nu^3}{3\gamma^2\epsilon}\kappa^4\right]^{-\frac{4}{3}}. \quad (6.6.15)$$

We note that when $\kappa \ll \left(\frac{3\gamma^2\epsilon}{8\nu^3}\right)^{\frac{1}{4}} \left(= \left(\frac{3\gamma^2}{8}\right)^{\frac{1}{4}} \kappa_d\right)$, the spectrum reduces to the expected form (6.6.9) for the inertial subrange. At the other extreme, when $\kappa \gg \left(\frac{3\gamma^2\epsilon}{8\nu^3}\right)^{\frac{1}{4}}$, the spectrum reduces, in agreement with (6.6.10), to

$$E(\kappa) \sim \left(\frac{\gamma\epsilon}{2\nu^2}\right)^2 \kappa^{-7}. \quad (6.6.16.)$$

The dissipation integral $2\nu\int_0^\infty \kappa^2 E(\kappa)\,d\kappa$ is determined principally by the transition range of values of κ lying in the neighbourhood of $\kappa = \left(\frac{3\gamma^2\epsilon}{8\nu^3}\right)^{\frac{1}{4}}$; provided γ is of order unity, as is implied by the physical basis of Heisenberg's transfer expression, we have here a confirmation of the interpretation (in § 6.4) that the wave-number $\kappa_d = 1/\eta = (\epsilon/\nu^3)^{\frac{1}{4}}$ marks the location of the range of strong viscous forces.

Measurements of the spectrum function $E(\kappa)$ (more precisely, of the one-dimensional spectrum function $\phi(\kappa)$, from which $E(\kappa)$ can be obtained by means of (3.4.18)) at very large wave-numbers have been made by Stewart and Townsend (1951) for the turbulence generated by a square-mesh grid, at Reynolds numbers of the grid up to $UM/\nu = 10^4$. These Reynolds numbers are not sufficiently high for the energy-containing range and the dissipation range of wave-numbers not to overlap, so that an equilibrium range, as we have defined it, does not exist under these conditions. Our hypotheses do not allow us to assume that the values of $E(\kappa)$ in the neighbourhood of $\kappa = \kappa_d$ are here independent of the energy-containing range of wave-numbers,† so that a comparison with the theoretical expression (6.6.15) is not wholly decisive. It was found experimentally that the value of $\frac{\kappa}{E}\frac{dE}{d\kappa}$ fell monotonically‡ at large values of κ, and at the highest wave-number covered by the measurements, viz. $\kappa = 1{\cdot}6\kappa_d$, it was about -10. For no choice of γ is it possible to make the graph $\epsilon^{-\frac{1}{4}}\nu^{-\frac{5}{4}}\left(\frac{\kappa}{\kappa_d}\right)^6 \phi(\kappa)$ against $\frac{\kappa}{\kappa_d}$, as predicted by (6.6.15), bear even an approximate resemblance to the experimental curve (which is shown in fig. 7.9). The measurements thus do not provide support for Heisenberg's expression for the energy transfer. A model of the turbulence which gives a better representation of the measured spectrum at these very large wave-numbers will be described in §7.4.

Another postulate about the transfer of energy which has an intuitive physical basis is that put forward by A. M. Obukhoff (1941). Obukhoff suggests that the transfer of energy across the wave-number magnitude κ can be considered as the consequence of a Reynolds stress, produced by wave-numbers greater than κ (the micro-component of the turbulence), acting on a mean rate of shear produced by wave-numbers less than κ (the macro-component). A natural assumption about the effective Reynolds stress produced by the micro-component is that it is proportional to the energy associated with wave-number magnitudes greater than κ, i.e. to $\int_\kappa^\infty E(\kappa')\,d\kappa'$. The contribution to the mean-square rate of strain from wave-number magnitudes less than κ is $2\int_0^\kappa \kappa''^2 E(\kappa'')\,d\kappa''$, so that provided we can use the root-mean-square rate of strain as a measure of the effective mean (in the sense of an average over the

† In fact, the measured spectrum curves E against κ/κ_d had the same shape at values of κ/κ_d above about 0·6, for several times of decay and several grid Reynolds numbers. The reasons, which lie outside the universal equilibrium theory and will be considered in the next chapter (§ 7.4), can be used to justify the comparison between the theoretical prediction (6.6.15) and measurements at large values of κ at Reynolds numbers which are not large enough for the universal equilibrium to exist.

‡ The same tendency can be detected in some measurements of the spectrum function in turbulent flow between two parallel planes made by J. Laufer (1950).

micro-component) rate of strain on which the Reynolds stress acts, the rate of transfer of energy becomes

$$S(\kappa) = \gamma \int_{\kappa}^{\infty} E(\kappa')\,d\kappa' \left[2\int_{0}^{\kappa} \kappa''^2 E(\kappa'')\,d\kappa'' \right]^{\frac{1}{2}}, \qquad (6.6.17)$$

where γ is an absolute constant. The equation (6.6.6) for the energy spectrum function can then be solved by elementary methods although the solution has not been given in closed form.

A criticism of the expression (6.6.17) is that it does not conform to the exact expression for the energy transfer as a double integral (see (6.6.4)), presumably because of the use of the root-mean-square rate of strain as an approximation to the effective mean rate of strain experienced by the micro-component. As with Heisenberg's expression for the transfer, the underlying physical idea seems to be suited more to the exchange of energy between distant wave-numbers than to the more important case of exchange between wave-numbers of the same order of magnitude. It is also clear from the calculated spectrum that something is amiss with (6.6.17). The function $E(\kappa)$ is found to be proportional to $\kappa^{-\frac{5}{3}}$ in the inertial subrange, as expected, and thereafter rises above the continuation of this $\kappa^{-\frac{5}{3}}$-curve until a value of κ is reached at which the dissipation integral $2\nu \int_{0}^{\kappa} \kappa''^2 E(\kappa'')\,d\kappa''$ has the assumed value ϵ. Evidently the spectrum function must be assumed to fall discontinuously to zero at this value of κ. None of these properties of the spectrum function is found in the measurements; nor does it seem likely, from their unreal character, that they will be.

It is clear that the rate of transfer of energy across wave-number κ is not a simple quantity, and it is unlikely that any simple physical postulate like the above two examples, or like that suggested by L. Kovasznay (1948), will describe it adequately. A proper solution of the problem is awaited with great interest.

CHAPTER VII

DECAY OF THE ENERGY-CONTAINING EDDIES

The arguments used in the first five chapters are deductive on the whole; they proceed from the basic equations governing the motion of the fluid to conclusions which, directly or indirectly, may be tested experimentally. However, in Chapter VI it became necessary to reverse the order of the reasoning in part, and to supplement the deductions from the basic equations with inferences from measurements; by arguing both forwards and backwards it was possible to obtain an understanding of one aspect of the turbulence which, while it may later be found inadequate in one or two minor respects, is unlikely to be wrong in general principle. In this chapter the process of reversing the order of reasoning is carried still further, and we shall be obliged to rely to a considerable extent on the empirical evidence. As a consequence the certainty of our interpretation of the measurements is much less and the theoretical ideas described below still have the status of intuitive hypotheses only.

7.1. The decay of total energy

One of the first measurements to be made in the turbulent motion behind grids placed in a uniform stream was the decay of the total kinetic energy of the turbulence. If the grid is composed of bars or rods spaced regularly at a distance M apart in a square array, it is found that $\overline{u_1^2}$, $\overline{u_2^2}$ and $\overline{u_3^2}$ become independent of position across the stream at a distance of about $10M$ from the grid; our theory is therefore concerned with measurements at greater distances than this. It is also found that $\overline{u_1^2}$, $\overline{u_2^2}$ and $\overline{u_3^2}$ are approximately equal at this position, and remain so until the permanent anisotropy of the biggest eddies becomes important in the final period of the decay. The determination of the energy decay curve thus requires only a single measurement at various stages of the decay, that is, at different distances downstream from the grid. Experimentally, the velocity component parallel to the mean flow is the easiest to measure with a hot-wire anemometer, and it is to this component that the following measurements relate, even though the symbol to be used to denote the root-mean-square (viz. u) suggests that it is an arbitrary velocity component.

The earliest measurements of u at different stages of the decay were a little confused by inadequate corrections for the errors introduced by the finite length of the recording hot wire, by the existence in the wind-tunnel stream of turbulence from sources other than the grid, and by the inclusion of observations at positions in the range $0 < x < 10M$. In recent years the measurements have been considerably refined, and there is now general agreement about the dependence of u on distance from the grid. It has been found that the energy decays according to the simple law

$$u^{-2} \propto (x - x_0), \qquad (7.1.1)$$

for a certain range of values of the distance x downstream from the grid, x_0 being a constant, while for larger values of x the energy decreases more rapidly and ultimately obeys the decay law (5.5.7) already established for the final period of the decay. The period of validity of (7.1.1) will be termed the *initial period* of the decay.

The law (7.1.1) has been established for a wide range of Reynolds number of the grid (measured by UM/ν) and for various shapes of grid, and it is possible that it is universally valid for homogeneous turbulence. As a sample of the measurements, fig. 7.1 shows the variation of u downstream from three grids of different shape; the first was composed of circular rods of diameter $M/5{\cdot}33$, spaced a distance M between centres and lying in two perpendicular directions (this is the type of grid normally used) (Batchelor and Townsend, 1948*a*), the second was composed of circular rods of diameter $M/2{\cdot}68$ spaced a distance M between centres and lying in one direction only, and the third was like the second but composed of slats of rectangular section $1{\cdot}95M$ by $0{\cdot}21M$ (Stewart and Townsend, 1951). In each case the relation (7.1.1) is satisfied, the only variation in the three cases being in the constant of proportionality (apart from small changes in x_0).

If measurements are made with grids of the same geometrical shape, it is found that the primary effect of using different values of U (the stream speed) and of M is to vary the effective units of u and x.

The relation (7.1.1) can thus be written as

$$\frac{U^2}{u^2} = a\left(\frac{x}{M} - \frac{x_0}{M}\right), \tag{7.1.2}$$

where a depends principally on the shape of the grid and may vary slowly with the grid Reynolds number UM/ν. Part of the dependence of a on grid shape and Reynolds number is undoubtedly accounted for by writing

$$a \propto 1/C_D,$$

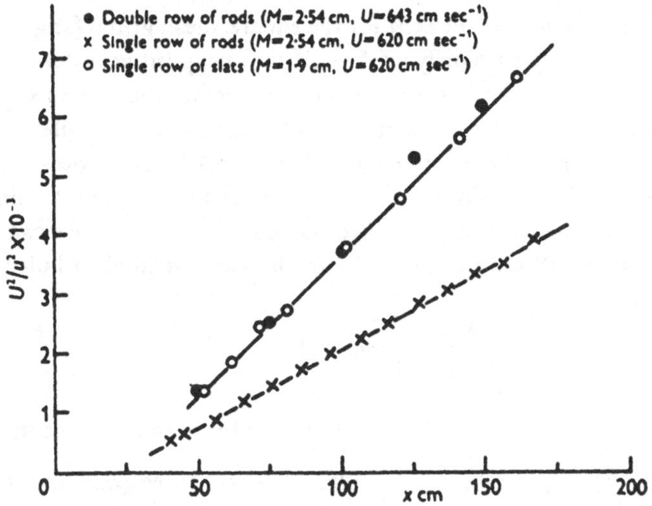

Fig. 7.1. Decay of energy behind different grids (after Batchelor and Townsend, 1948*a*; and Stewart and Townsend, 1951).

where $C_D . \frac{1}{2}\rho U^2$ is the force which unit area of the grid exerts on the stream in order to create the energy of the turbulence. In addition, the geometry of the grid must determine the effective unit of length for the turbulence. If we write

$$\frac{U^2}{u^2} = \frac{b}{C_D}\left(\frac{x}{M} - \frac{x_0}{M}\right), \tag{7.1.3}$$

the values of b for the three grids described above are found to be 101 (double row of rods), 53 (single row of rods), and 91 (single row of slats), showing that if the effective unit of the distance x for the double row of rods is taken as M, the effective units for the other two grids are $1{\cdot}91M$ and $1{\cdot}11M$ respectively, where M is in each case the periodic interval; these figures are consistent with the notion that the effective unit of length is measured by the average size of the energy-containing eddies generated by the grid and that the latter are determined roughly by the size (both across and parallel to the stream) of the holes in the grid. For the commonly used square-mesh grid the constant a in (7.1.2) has a value near 134 and x_0/M is about 10.†

The constant x_0 in (7.1.3) determines the position of the virtual origin at which the energy would be infinite, and is found to lie between about $5M$ and $15M$ for the square-mesh grid (cf. fig. 7.1), the value varying slowly with UM/ν. The positive value of x_0 reflects the fact that a certain time elapses before the motion settles down to the high rate of dissipation characteristic of turbulence.

It has already been remarked (see (5.5.7)) that a power-law variation of u^2 has simple consequences for the quantity λ. Since the time of decay in the idealized homogeneous turbulence can be identified with the quantity x/U in the experimental turbulence, we have

$$\lambda^2 = -\frac{10\nu u^2}{U\dfrac{\mathrm{d}u^2}{\mathrm{d}x}} = \frac{10\nu}{U}(x - x_0) \tag{7.1.4}$$

from (7.1.1). Values of λ^2 may be obtained from measurements of, say, $\overline{\left(\dfrac{\partial u_1}{\partial x_1}\right)^2} = \dfrac{u^2}{\lambda^2}$, and values so obtained for the three grids described above are shown in fig. 7.2. Lines of slope $10\nu/U$ are also drawn in the figure and the agreement with (7.1.4) demonstrates the general accuracy of the measurements. The agreement also confirms, in part, the assumptions that the experimental turbulence is homogeneous and that the Navier-Stokes equation (on which (7.1.4) is based) is valid. We note the very interesting consequence of (7.1.1)

† These data supply the following estimate of the length η (defined in (6.4.1)) during the initial period:

$$\eta = \left(\frac{\nu^3}{\epsilon}\right)^{\frac{1}{4}} \approx 3{\cdot}1M\left(\frac{x}{M}\right)^{\frac{1}{2}}\left(\frac{UM}{\nu}\right)^{-\frac{3}{4}}.$$

Taking $x/M = 20$ and $U = 2500$ cm sec.$^{-1}$, we find that the smallest value of η likely to be encountered in wind-tunnel practice is of the order of $0{\cdot}01M$. Since η is a measure of the order of magnitude of the effective lower bound of the range of eddy sizes, it is quite clear that the molecular structure of the medium is irrelevant to the turbulent motion, as was claimed in § 1.2.

that the Reynolds numbers $u\lambda/\nu$ and (using (6.1.1)) ul/ν remain constant during the initial period of decay. If the law of decay is represented by $u^{-2} \propto (x-x_0)^n$, $u\lambda/\nu$ and ul/ν decrease during decay for $n>1$ and increase for $n<1$.

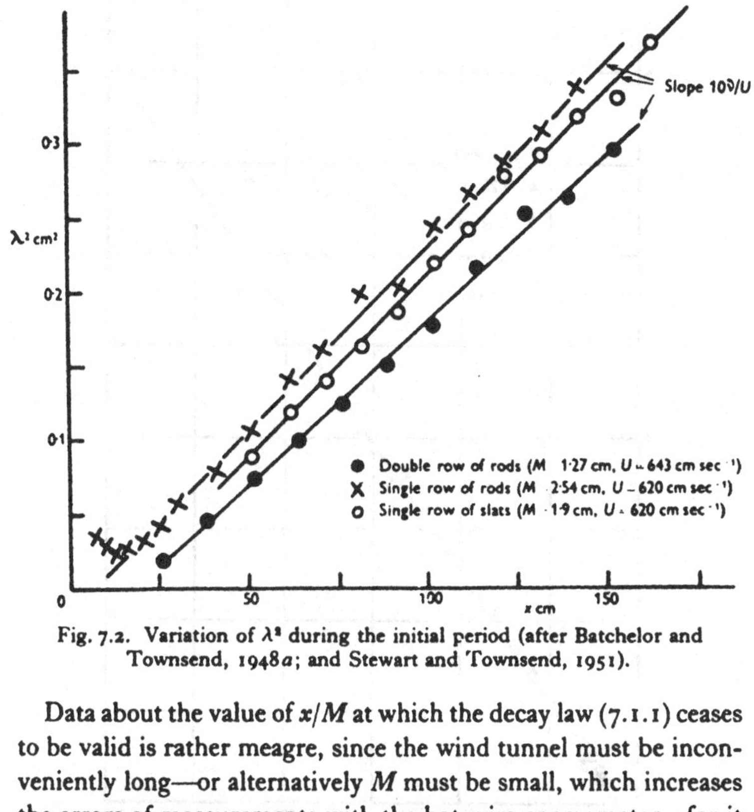

Fig. 7.2. Variation of λ^2 during the initial period (after Batchelor and Townsend, 1948*a*; and Stewart and Townsend, 1951).

Data about the value of x/M at which the decay law (7.1.1) ceases to be valid is rather meagre, since the wind tunnel must be inconveniently long—or alternatively M must be small, which increases the errors of measurements with the hot-wire anemometer—for it to be observable. The available estimates (Batchelor and Townsend, 1948*a*) all refer to square-mesh grids and indicate that the duration of the initial period (measured non-dimensionally) does not vary greatly with the Reynolds number UM/ν. The energy begins to decay more rapidly than is predicted by the relation (7.1.1) at a value of x/M which seems to lie between about 120 and 200, being greater for higher Reynolds numbers; fig. 7.3 shows a typical set of energy measurements (behind a square-mesh grid) extending beyond the initial period of decay (Batchelor and Townsend, 1948*a*).

The length of the transitional period between the initial and final periods of decay depends markedly on the Reynolds number, as might be expected, since the final period cannot occur until

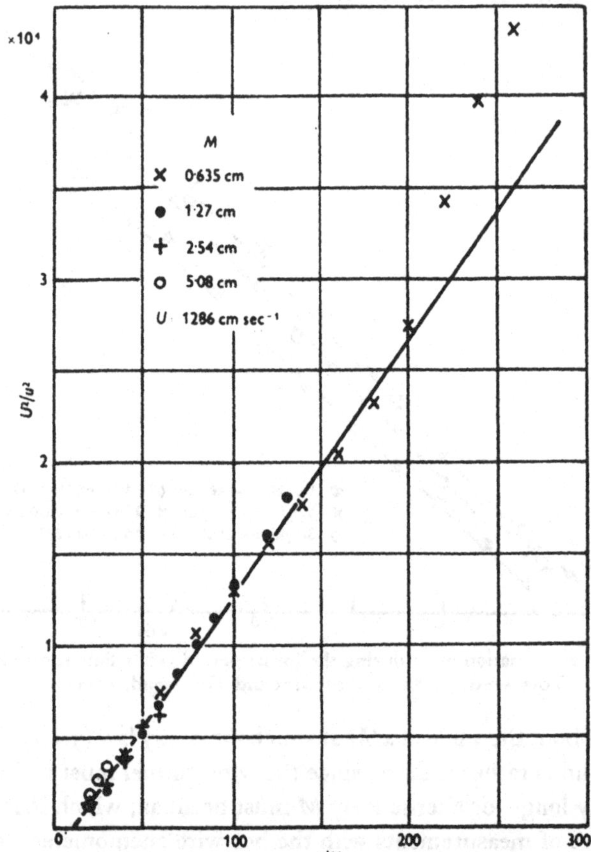

Fig. 7.3. Decay of energy during and beyond the initial period (after Batchelor and Townsend, 1948*a*).

the absolute energy u^2 (in contradistinction to the relative energy u^2/U^2) has fallen to a very low level. When UM/ν is as low as 650, it has been found (see § 5.4) that the final period of decay begins at about $x/M = 400$, but at appreciably higher Reynolds numbers the final period of decay is quite out of reach of experiments in a wind tunnel.

These strikingly simple facts about the decay of energy in the initial period seem to call for a correspondingly simple explanation. At the time of writing there is no explanation which is entirely satisfactory, the difficulty with the suggested explanations being that they predict too much, as we shall see in the following sections.

7.2. Evidence for the existence of a unique statistical state of the energy-containing eddies

The observation of an energy-decay law which has the same form for different grid Reynolds numbers and for grids of different shape suggests that the turbulence rapidly settles down to a state which is independent—partly, if not wholly—of the conditions of its formation. The same suggestion arises from the approximate isotropy of the turbulence at an early stage of the decay, and from the fact that the decay law is such as to leave the Reynolds number of the energy-containing eddies constant during decay. We should be tempted, in fact, to suppose that all the energy-containing eddies rapidly adjust themselves to some stable statistical state of which the decay law (7.1.1) is a particular manifestation. However, we must bear in mind the warning of § 6.1, where it was pointed out that if the characteristic time of the energy-containing eddies (considered as a whole) is measured by l/u, the decay time of the turbulence is of the same order, suggesting that there is insufficient time available for an adjustment to a stable statistical state; tempting though the supposition might be, it must clearly be approached with caution. Perhaps, at the least, we can expect to find that there is a continuous gradation from absolute equilibrium at the very largest wave-numbers, through a kind of semi- or quasi-equilibrium at intermediate wave-numbers, to a state of direct dependence on the initial conditions at the lowest wave-numbers. The two ends of this range have already been examined (in Chapter VI and § 5.3 respectively), and we must now consider the intermediate wave-numbers. The development of some kind of asymptotic statistical state will be manifested as a trend towards a simple form of the spectrum and correlation functions during decay—perhaps one that involves no change in shape—so that it is to the experimental evidence concerning these functions that we must look.

In order to determine whether a statistical state of the energy-containing eddies which is independent of the initial conditions is set up during the initial period of decay, Stewart and Townsend (1951)† have measured the two transverse velocity correlations, $R_{11}(0, r, 0)$ and $R_{11}(0, 0, r)$, for the three different grid shapes described in the previous section at the same stage of decay (the x_1-axis lies downstream, and the x_2- and x_3-axes are parallel and perpendicular respectively to the rods making up the grid). Fig. 7.4(*a*) shows a comparison of the correlation functions for the double row of circular rods and the single row of circular rods, the values of M for the two grids being so chosen that they gave approximately the same turbulent energy at the same distance downstream, and fig. 7.4(*b*) shows a similar comparison for the double row of circular rods and the single row of rectangular slats. (It was found that $\overline{u_1^2} = \overline{u_2^2} = \overline{u_3^2}$ for each grid at the positions of the measurements.) The measurements show that for values of r which are not too large the transverse correlations are the same for all three grids, and the transverse correlations in different directions are identical for the two grids consisting of a single row of rods. Thus for these values of r, the velocity correlation has an isotropic form which seems to be independent of the geometry of the grid. For larger values of r the correlation functions are not the same for the three grids, nor do they have an isotropic form for the two single-row grids.‡ These differences at large values of r are found to persist throughout the initial period. Stewart and Townsend have calculated the spectrum function $E(\kappa)$ from the measured correlation functions (on the assumption that the turbulence is isotropic) and find that the range of wave-numbers affected by the geometry of the grid contains about one-fifth of the total energy.

These experiments suggest that the turbulence very quickly settles down to a statistical state in which the distribution of about 80% of the total energy, distributed over the high wave-number components, is approximately independent of the initial conditions.

† This paper is an account of a comprehensive set of measurements designed to test the ideas considered in this and the preceding chapter, and as a consequence we shall need to refer to it often.

‡ Lack of isotropy at large values of r for the double row of rods is not revealed by a comparison of $R_{11}(0, r, 0)$ and $R_{11}(0, 0, r)$, but it does exist, as was seen from the measurements described in § 5.4.

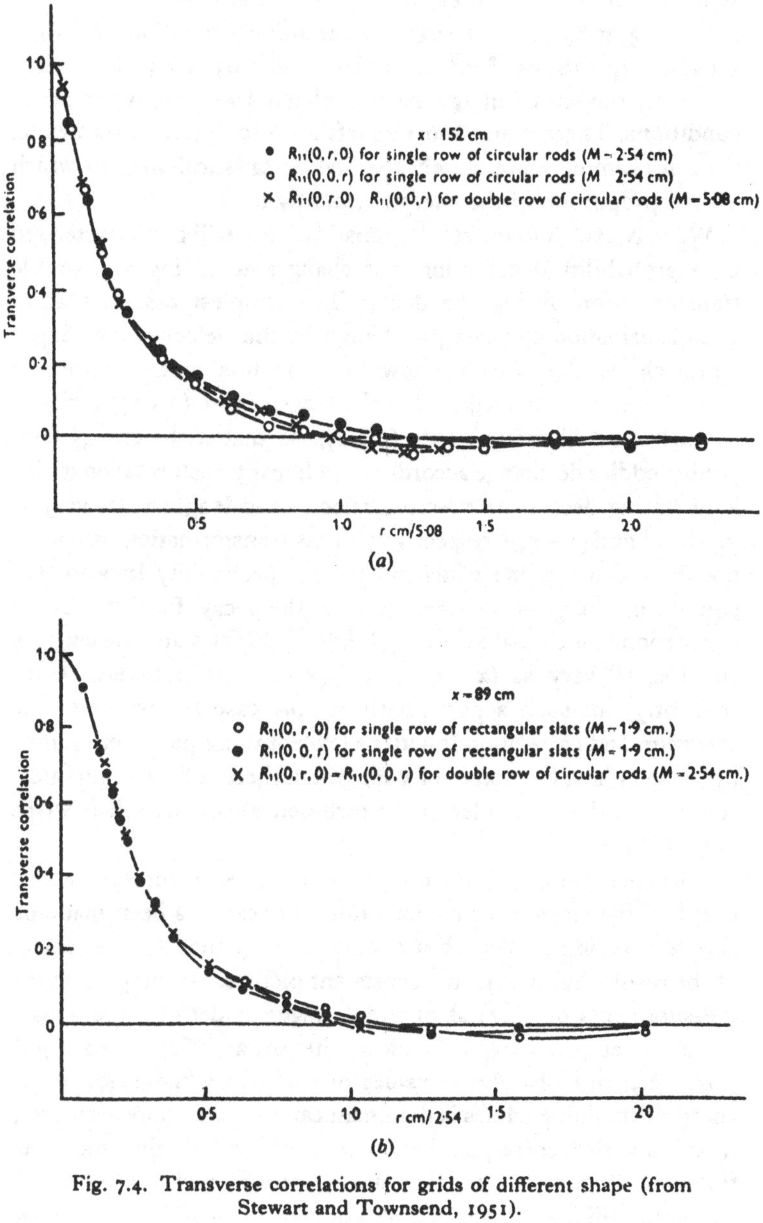

Fig. 7.4. Transverse correlations for grids of different shape (from Stewart and Townsend, 1951).

It is not clear, at the present time, why the dependence of the remaining 20% of the energy on the initial conditions does not cause the quantities $\overline{u_1^2}$, $\overline{u_2^2}$ and $\overline{u_3^2}$ to be slightly unequal, nor is it clear why the law of energy decay is affected so little by the initial conditions. These points must be left open for a time; meanwhile, we can examine more closely the stable statistical state to which a large proportion of the energy evidently tends.

We may expect that a stable statistical state will be characterized by a probability distribution that changes according to a simple transformation during the decay. The simplest case is that of a transformation involving a change in the velocity and length scales alone. We already know that the total energy varies as $(x-x_0)^{-1}$ and consequently that the length l (see (6.1.1)) varies as $(x-x_0)^{\frac{1}{2}}$ (and likewise λ—see (7.1.4)), so that if the energy-containing eddies do change according to a linear transformation of this kind during decay, the effective velocity and length units vary as $(x-x_0)^{-\frac{1}{2}}$ and $(x-x_0)^{\frac{1}{2}}$ respectively. This transformation, it will be noted, is also the one which relates the probability laws for the equilibrium range at different stages of the decay, for if the rate of dissipation is such that $\epsilon \propto (x-x_0)^{-2}$, the velocity v and the length η (see (6.4.1)) vary as $(x-x_0)^{-\frac{1}{2}}$ and $(x-x_0)^{\frac{1}{2}}$ respectively. Consequently, for such a particularly simple case the whole of the spectrum and correlation functions (except those parts determined by the very smallest values of κ for which there is little or no interaction with the remainder of the turbulence) preserve their shape during decay.

An experimental examination of the degree of preservation of shape of the statistical functions during decay has been made on several occasions, although it is only recently that the assessment of the results has provided a consistent picture. We may quote the measurements of $u^2 f(r)$ at different stages of decay, for a square mesh grid at $UM/\nu = 5300$, made by Stewart and Townsend (1951) (measurements of $u^2 f(r)$ at values of r at which the effect of the energy-containing eddies is dominant can be made more accurately than can the corresponding measurements of the spectrum function). Fig. 7.5 shows the results plotted in such a way that the curves for different values of x/M would coincide if the only change during decay were a change in the effective velocity and length units. Near $r = 0$ the curves necessarily coincide in order to conform to the exact relation (3.4.7). It will be seen that at larger values of r the use of r/λ as independent variable has clearly brought the curves closer to coincidence, but that there are still significant differences between the curves.

The meaning of the differences is perceived more readily from the set of three-dimensional spectrum curves calculated from the

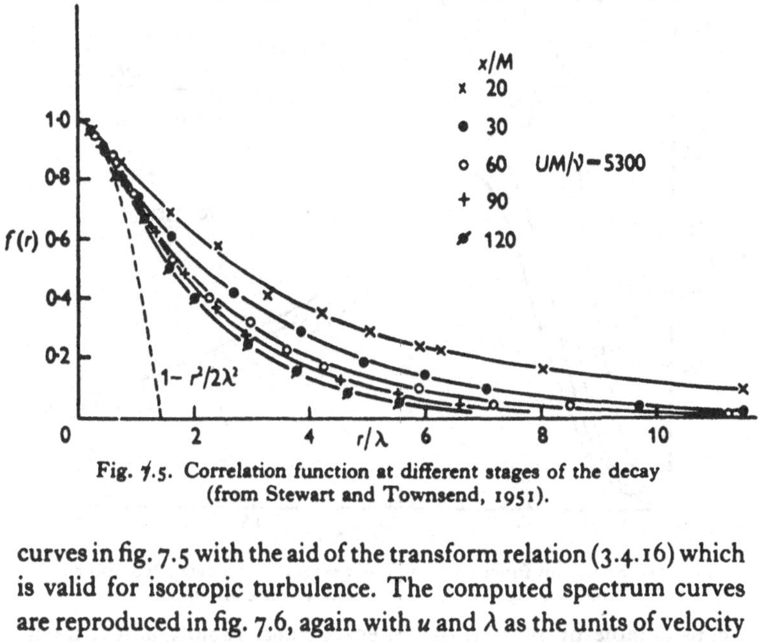

Fig. 7.5. Correlation function at different stages of the decay (from Stewart and Townsend, 1951).

curves in fig. 7.5 with the aid of the transform relation (3.4.16) which is valid for isotropic turbulence. The computed spectrum curves are reproduced in fig. 7.6, again with u and λ as the units of velocity and length. The curves are now approximately coincident for $\kappa\lambda > 1$, the differences being concentrated in the range of smaller wave-numbers. This latter range, for which the spectrum does not preserve its shape during decay, contains an appreciable fraction of the total energy—of the order of 25%—and is also the range for which the motion is not isotropic in the case of grids composed of rods extending in one lateral direction only, as already described.† The measurements thus support the theoretical expectation that

† The computed curves in fig. 7.6 cannot be used to confirm the theoretical deduction that $E(\kappa) \approx C\kappa^4$ (where C is constant) at very small values of κ, because the assumption of isotropy on which the relation between figs. 7.5 and 7.6 is based is not accurate at these small wave-numbers.

some at least of the energy-containing eddies, viz. those corresponding to the smaller values of κ, are dependent on the initial conditions throughout the initial period of decay. In view of this dependence on the shape of the grid, it seems unlikely that an analytical deduction of the statistical properties of the motion associated with these small wave-numbers could be given (except for the continuity relations, valid near $\kappa = 0$, derived in § 5.3) or that it would have general validity. The measurements also confirm that

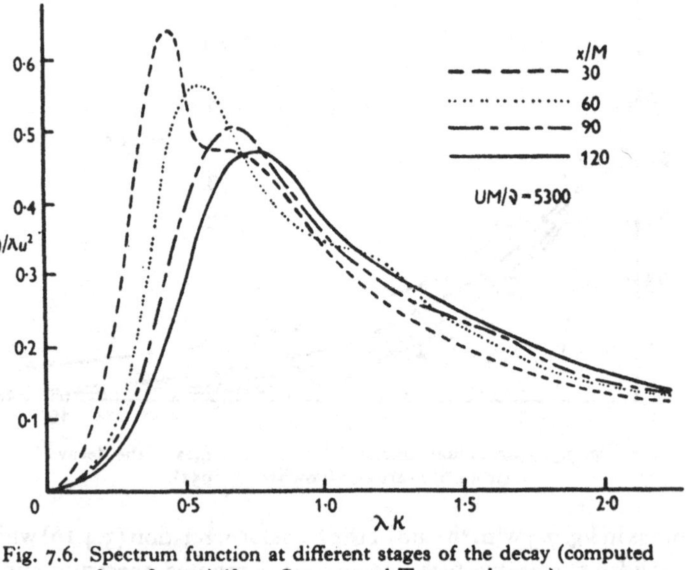

Fig. 7.6. Spectrum function at different stages of the decay (computed from fig. 7.5) (from Stewart and Townsend, 1951).

a considerable proportion of the energy *does* rapidly adjust itself to a statistical state which is approximately independent of the initial conditions and, also, apart from a linear transformation, independent of the time of decay, and this part of the motion is a fit subject for a general theory.

It is at first sight surprising that the stable statistical state which a large proportion of the energy rapidly attains is such that the variation with time of decay is equivalent to variation of the effective velocity and length units only, since this is the degree of simplicity attained in the equilibrium range. Part of the explanation is that although further new parameters (evidently one only) are needed to specify the motion associated with wave-numbers too small to lie in the equilibrium range, the viscosity is not a relevant parameter for these wave-numbers, so that there is both a loss and a gain in simplicity by comparison with the equilibrium range. The remainder of the explanation, viz. why only one new parameter—u and l in place of ϵ ($\propto u^3/l$)—should be needed to take account of the lack of equilibrium of the energy-containing eddies is not yet clear.

The situation seems to be that at very large wave-numbers the statistical functions vary only inasmuch as the velocity $v = (\nu\epsilon)^{\frac{1}{4}}$ and the length $\eta = (\nu^3/\epsilon)^{\frac{1}{4}}$ vary,† while for a range of smaller wave-numbers the statistical functions vary only inasmuch as u and l vary. As we move out of the former range, in the direction of decreasing wave-numbers, the parameter ν becomes less relevant while u and l become relevant separately and not simply in the combination u^3/l ($\propto \epsilon$). For either of the two ranges separately, the spectrum and correlation functions measured under different conditions can be brought into coincidence by a linear transformation. Also, functions measured at different stages of the decay can be brought into coincidence over the two ranges together by a linear transformation, because v/u and η/l remain constant during the initial period of the decay. However, functions measured at different Reynolds numbers—that is, in effect, for different values of ν—cannot be made coincident over the two ranges together, but will form a singly-infinite family of curves of different shape, one for each value of the Reynolds number ul/ν (or, equivalently, for each value of v/u or of η/l) which remains constant during decay.

Further evidence in support of this empirical picture is provided by the data shown in figs. 7.7 and 7.8 and taken from the paper by Stewart and Townsend (1951). Fig. 7.7 shows measurements of the longitudinal one-dimensional spectrum $\phi(\kappa)$ (see (3.4.17)) at different stages of the decay of turbulence generated by a square-mesh grid, and at different values of the Reynolds number UM/ν. The measurements are plotted with v and η as effective velocity and length units and show the approximate preservation of shape during decay that has already been noted. (The lack of preservation of shape at small wave-numbers is obscured in these measurements

† We have not yet shown that these are the determining parameters when the Reynolds number is not large enough for the existence of a universal equilibrium of the kind described in the preceding chapter, but relevant evidence will be presented later.

by the scatter of the observations in this range.) The curves corresponding to different Reynolds numbers are quite clearly not brought into coincidence over the range of κ in which most of the energy lies. At higher values of κ, however, the curves are more

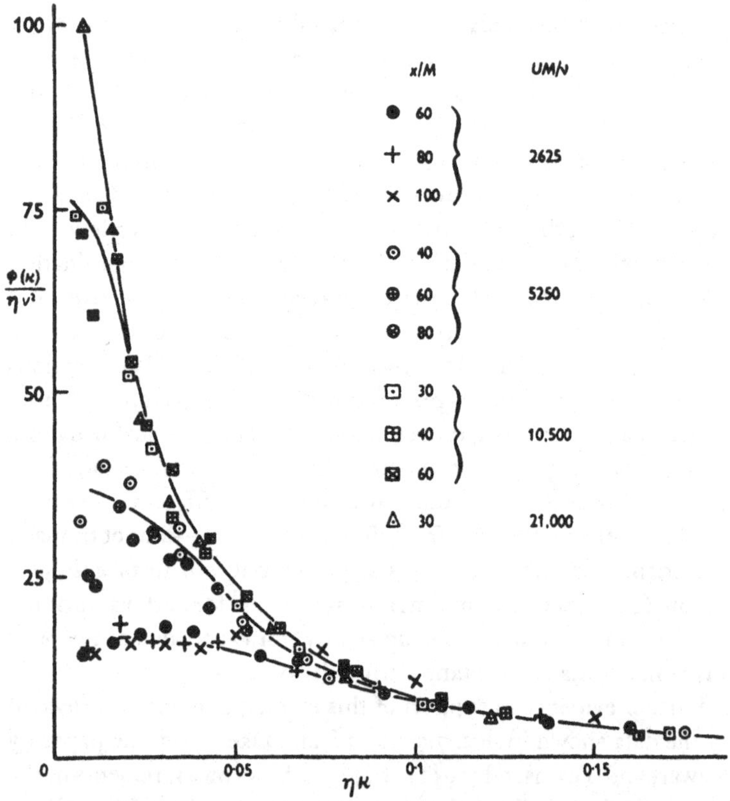

Fig. 7.7. Spectrum functions (one-dimensional) at different decay times and different Reynolds numbers (from Stewart and Townsend, 1951).

nearly coincident, and Stewart and Townsend describe further measurements which show that the coincidence becomes increasingly good as κ is increased still further. (This evident universality of the spectrum shape at large values of κ will need some further consideration (see § 7.4), since the Reynolds numbers are not large enough for a universal equilibrium of the kind described in Chapter VI to exist.)

Fig. 7.8 shows, on the other hand, how the statistical functions may be brought into complete coincidence over most of the energy-containing range, provided the Reynolds number is not too small, when u and l are used as units. For experimental reasons correlation functions are more suitable for an examination of the energy-containing eddies, and fig. 7.8 therefore shows measurements of the longitudinal correlation $u^2 f(r)$ for a square-mesh grid at different Reynolds numbers but at the same value of x/M in order not to have the diagram confused by the lack of similarity, during decay, of the

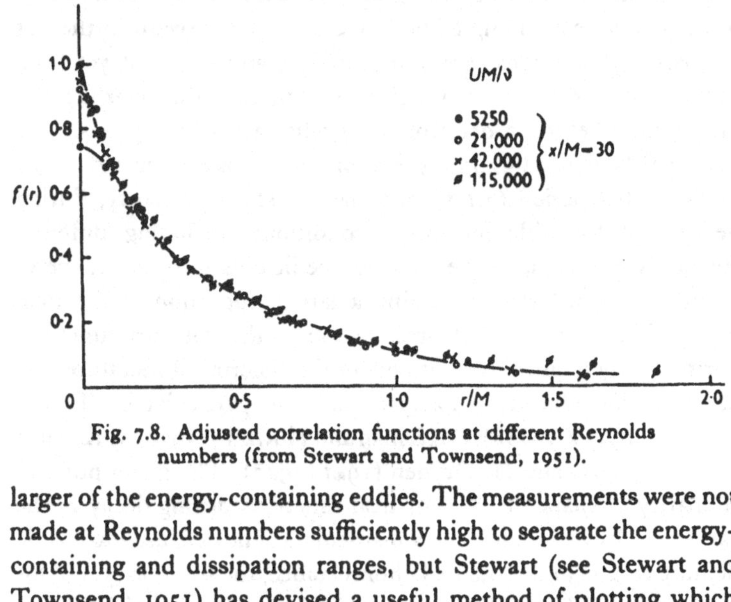

Fig. 7.8. Adjusted correlation functions at different Reynolds numbers (from Stewart and Townsend, 1951).

larger of the energy-containing eddies. The measurements were not made at Reynolds numbers sufficiently high to separate the energy-containing and dissipation ranges, but Stewart (see Stewart and Townsend, 1951) has devised a useful method of plotting which corrects for this departure from the desired conditions. The method consists of using as the unit of velocity, not the measured u, but $u(1+\Delta)^{\frac{1}{2}}$, where Δ is the (small) fraction by which the measured energy falls short of the amount which would be present if viscous forces acted only at infinitely large wave-numbers. There is no means of estimating Δ theoretically, but since it vanishes with ν we may suppose that to the first approximation it is proportional to $(UM/\nu)^{-1}$. Stewart and Townsend made the equivalent assumption that

$$\Delta = \alpha \frac{\lambda^2}{M^2}, \qquad (7.2.1)$$

and found that the application of this correction, with $\alpha = 6{\cdot}1$, brought all the measured curves into approximate coincidence except at large and at small values of r, as shown in fig. 7.8. (The unit of length in the figure is M, to which l is proportional at a fixed stage of the decay.) The coincident part of the curves is probably a good approximation to the asymptotic form of the correlation function at infinitely large Reynolds number.

7·3. The quasi-equilibrium hypothesis

Some theoretical ideas which help to 'explain' the observations of the energy containing eddies have been put forward in the last few years. These ideas are still tentative, and have not yet been thoroughly clarified, so that their proper place in this book is after, rather than before, a description of what actually happens. The principal facts needing interpretation are the law of energy decay in the initial period (7.1.1), and the statistical similarity, during decay and for different initial conditions, including different Reynolds numbers, of the Fourier coefficients over the range of (high) wave-numbers containing a large proportion of the total energy. The notion of 'self-preservation' of the statistical functions during decay is of course not new in the theoretical literature and has been introduced by many authors, in particular by T. von Kármán (1937*a*, 1948*a*), v. Kármán and Howarth (1938), v. Kármán and Lin (1951) and H. Dryden (1941, 1943). The assumption of similarity of shape of the statistical functions during decay in the earlier works was principally a mathematical device, used to enable definite results to be obtained. For instance, the solution (5.4.8) for the longitudinal correlation function in the final period of decay was given in this way (for isotropic turbulence) by v. Kármán and Howarth (1938) as one of an infinite family of self-preserving solutions. To find such solutions has been one task; to determine the conditions under which they can and do provide a correct description of the turbulence is another. It is this latter task which has engaged much attention in the last five years, but even so most of the established results are negative, and our positive results still rest insecurely on vague intuitive arguments (vague for most of us—clear and precise for the inspired few!).

The principal contribution to the understanding of the observed similarity during decay has been made by Heisenberg (1948*b*); he framed his arguments in terms of the special form of the energy transfer described in §6.6, but in their general aspects they are essentially similar to those used hereunder. We have seen that the hypothesis of a universal equilibrium of the motion associated with large wave-numbers is self-consistent and plausible, and finds some experimental support. Admitting the validity of this hypothesis, we can go on to consider the nature of the motion at wave-numbers a little lower than those in the equilibrium range. The properties of these components of the turbulence are affected directly by the time-dependence of the turbulence as a whole, in such a way, we may expect, as to convert the absolute equilibrium which exists at higher wave-numbers into a moving or quasi-equilibrium. A quasi-equilibrium of a system of interacting degrees of freedom is an asymptotic state which is as near to equilibrium as is consistent with a finite rate of change of total energy of the system. The simplest quasi-equilibrium, that is, the kind that will occur at wave-numbers close to, but outside, the equilibrium range, will require for its statistical specification one parameter extra to those needed for the equilibrium range; this parameter will presumably describe in some way the stage of the decay of the turbulence. The division between the ranges of equilibrium and quasi-equilibrium is not sharp, of course, and as suggested earlier there will be a continuous variation from conditions of absolute equilibrium at the highest wave-numbers to conditions of complete permanence at the lowest wave-numbers. The essential point of the argument is that the higher the wave-number of the component of the motion, the smaller is its characteristic reaction time; the reasons for this are admittedly not rigorous, and are based partly on the form of the equation of motion, which equates the time-derivative of the velocity to the sum of length-derivatives, and partly on casual visual studies of turbulent motion.

The choice of the extra parameter needed to specify mean values associated with the quasi-equilibrium is suggested strongly by known solutions of the hydrodynamical equations in cases of unsteady motion. When, in these latter problems, there is no length involved in the motion permanently (e.g. through the boundary conditions), it is frequently found that the solution tends asymptotically to a function of $\mathbf{x}/(\nu t)^{\frac{1}{2}}$ whatever the form of the initial conditions. Speaking very generally, the initial conditions produce 'free oscillations' which are soon damped out, and the remaining 'forced oscillations' are produced by the boundary conditions on $\mathbf{x}$; if these boundary conditions supply no natural length unit, $(\nu t)^{\frac{1}{2}}$ becomes the single length unit for the velocity distribution. Such a velocity distribution then preserves its shape during the decay. These similarity solutions occur chiefly in problems in which the non-linear term in the equation of motion is zero or does not play an important part (as, for instance, in the case of fluid bounded by an infinite plane in steady motion, parallel to itself, the initial distribution of velocity in the fluid being arbitrary); the new suggestion is that this type of asymptotic solution will occur in cases in which the non-linear term has no particular form but operates statistically to transfer energy between the different wave-number components. It has yet to be shown that the boundary conditions insert no length scale into the motion associated with the quasi-equilibrium, but this is certainly suggested by the observed decay law which, if extrapolated backwards, is consistent with a turbulent motion of infinite energy and *zero* length scale at the origin of time.

We are thus led to the following quasi-equilibrium hypothesis:

'When the Reynolds number is large enough for a universal equilibrium to exist at high wave-numbers,† the motion associated with this equilibrium range *and* the neighbouring range of smaller wave-numbers is uniquely determined statistically by the parameters ν, ϵ and t (t being measured from the virtual instant at which the turbulence has infinite energy).'

If our intuitive ideas about the quasi-equilibrium are correct, the influence of the parameter t will diminish as κ increases, giving a gradual transition to the conditions of the equilibrium range.

An immediate consequence of this hypothesis is that the dimensionless ratio $\epsilon t^2/\nu$ must be constant ($=R$, say) during the decay,

† This restriction on the Reynolds number is not made to ensure the existence of some kind of equilibrium and of some kind of quasi-equilibrium, but to ensure that their properties are relatively simple. Some relaxation of the condition on the Reynolds number could be made, as has been pointed out by S. Goldstein (1951); see also § 7.4.

for there is no way in which it could depend on t alone. Hence,

$$\frac{du^2}{dt} = -\frac{\frac{2}{3}R\nu}{t^2}, \tag{7.3.1}$$

where R is determined by the initial conditions of the turbulence. In particular, for turbulence generated by passing fluid at speed U through grids of similar shape and of characteristic length M, R must be determined by the Reynolds number UM/ν alone. The law of decay follows from (7.3.1) as

$$u^2 - u_1^2 = \frac{\frac{2}{3}R\nu}{t}, \tag{7.3.2}$$

where u_1 is an arbitrary constant. This decay law is consistent with the observed decay law (7.1.2) (assuming that the latter is equally valid at Reynolds numbers larger than those used in the experiments), when t is replaced by the equivalent quantity $(x-x_0)/U$, provided that $u_1^2 \ll u^2$.†

The physical interpretation of the constant u_1 is obscure. If the quasi-equilibrium is closed to the supply of energy from smaller wave-numbers (in which event it would certainly be easier to think about) we could regard $\frac{3}{2}u_1^2$ as the energy (per unit mass of fluid) which lies outside the quasi-equilibrium and which necessarily remains constant. On this interpretation, the energy of any uniform translation of the fluid which exists contributes to both $\frac{3}{2}u_1^2$ and $\frac{3}{2}u^2$; likewise Fourier components with very small wave-numbers have very little interaction with larger wave-numbers (see § 5.3), and their contribution to $\frac{3}{2}u_1^2$ would remain approximately constant. The empirical result that $u_1^2 \ll u^2$ seems then to suggest that all but a negligible fraction of the energy takes part in the quasi-equilibrium. However, we have seen that there is strong evidence from the measured spectrum and correlation functions that at least

† The meaning of the constant R emerges from a comparison of (7.3.2) and (7.1.2):

$$R = \frac{\epsilon t^2}{\nu} = 0{\cdot}15\left(\frac{u\lambda}{\nu}\right)^2 = \frac{3}{2a}\left(\frac{UM}{\nu}\right), \tag{7.3.3}$$

showing that R is a direct measure of the Reynolds number of the grid. For a square-mesh grid we have

$$R = \frac{1}{90}\left(\frac{UM}{\nu}\right) \approx 1{\cdot}4\left(\frac{ul}{\nu}\right),$$

if l is taken as the longitudinal scale L_p.

25%—certainly not a negligible fraction—of the energy is associated with motion which is dependent on the initial conditions and which is therefore excluded from a quasi-equilibrium. This illustrates the difficulty which faces theories of similarity during decay; the need for consistency with the observed decay law makes it difficult for them to avoid predicting too much.

However, it is not necessary to assume that the quasi-equilibrium is closed; indeed, it is difficult to see how it could be if it does not contain nearly all the energy of the turbulence. Both C. C. Lin (v. Kármán and Lin, 1951) and S. Goldstein (1951) regard as natural the assumption that the quasi-equilibrium or similarity includes all wave-numbers from which a contribution to ϵ, i.e. to $\int_0^\infty \kappa^2 E(\kappa)\, d\kappa$, is made, but may exclude a range of (smaller) wave-numbers from which an appreciable contribution to total energy, i.e. to $\int_0^\infty E(\kappa)\, d\kappa$, is made. (It should be noted that both authors have in mind Reynolds numbers which are not necessarily so high that a universal equilibrium exists, so that the dissipation and energy-containing ranges may overlap.) There are no mathematical inconsistencies in this picture, but the physical basis for the certain inclusion of the dissipation within the similarity range is not clear, and the observation that u_1 is effectively zero (during the initial period) is left unexplained. The closeness of the observed decay law to $u^2 \propto t^{-1}$ may be due to the fact that, given that a large proportion—of the order of three-quarters—of the energy is observed (see fig. 7.5) to lie within the range of similarity (for reasons yet to be found; the *a priori* estimate would have been smaller) and decays as t^{-1}, the rate of decay of the remaining, non-similar, portion of the energy could vary a little with the initial conditions without causing a detectable departure from the law $u^2 \propto t^{-1}$ for the total energy. A complete understanding of the matter must wait on future work.

The existence of a certain proportion of the energy outside the quasi-equilibrium gives us a clue to the reason for the termination of the decay law (7.1.1) after a certain time. If the quasi-equilibrium embraced all the energy, the similarity laws and the decay law 7.3.2) (with $u_1^2 \ll u^2$) would continue for indefinitely large values of t. The same result would hold if the quasi-equilibrium took energy from lower wave-number components at a rate roughly proportional to t^{-1}. However, this transfer across the boundary of the quasi-equilibrium is likely to occur at the required rate only if the distribution of energy outside the quasi-equilibrium is suitable, and it is a reasonable guess that the distribution cannot remain suitable for very long. The length unit of the motion associated with the quasi-equilibrium is increasing as $t^{\frac{1}{2}}$, while the larger length scales of the motion outside the quasi-equilibrium are increasing more slowly (we have seen that for sufficiently small values of κ there is no change of the spectrum with time, and the corresponding parts of the spectrum curves in fig. 7.6 would be made coincident if the abscissa were $t^{\frac{1}{4}}\kappa$ instead of $t^{\frac{1}{2}}\kappa$). A continuous adjustment at the boundary is thus necessary, and since the distribution of energy outside the quasi-equilibrium is not maintaining its shape during the decay there can be no permanent relation between the two regions.

Other consequences of the quasi-equilibrium hypothesis concern the changes in the spectrum and correlation functions during decay. For instance, the spectrum function $E(\kappa, t)$ must be of the form

$$E(\kappa, t) = v^2 \eta E_{qe}\left(\eta\kappa, \frac{\epsilon t^2}{\nu}\right), \tag{7.3.4}$$

where v and η have the same meanings as before (see (6.4.1)), and E_{qe} is a universal function of two variables; κ must be suitably limited at the lower end in order to lie within the quasi-equilibrium range but may take indefinitely large values. It has already been pointed out that $\epsilon t^2/\nu$ has the constant value R during decay and that for geometrically similar grids R is directly proportional to UM/ν. Hence (7.3.4) agrees with the observation that the spectrum makes a linear transformation over the equilibrium and quasi-equilibrium ranges together during decay, but has a different shape for different grid Reynolds numbers. Since $\eta \propto t^{\frac{1}{2}}$ and $v \propto t^{-\frac{1}{2}}$ when $\epsilon \propto t^{-2}$, the predicted change in the effective length and velocity units during decay is in accord with observation, and (7.3.4) can equally well be written as

$$E(\kappa, t) = u^2 \lambda E_{qe}\left(\lambda\kappa, \frac{UM}{\nu}\right) \quad \text{or} \quad u^2 \lambda E_{qe}\left(\lambda\kappa, \frac{u\lambda}{\nu}\right), \tag{7.3.5}$$

as was inferred from the experiments, or, equivalently, as

$$E(\kappa, t) = u^2 l E_{qe}\left(l\kappa, \frac{UM}{\nu}\right), \tag{7.3.6}$$

where l is a length characteristic of the energy-containing eddies which lie within the quasi-equilibrium range. The effect of viscosity in all these expressions is confined to the equilibrium range (and may, if the Reynolds number is large enough, be confined even more narrowly, as described in §6.5), so that provided $l\kappa$ is not $\gg 1$, (7.3.6) reduces to

$$E(\kappa, t) = u^2 l E_q(l\kappa). \tag{7.3.7}$$

In other words, the family of curves E_{qe} against $l\kappa$ obtained for different values of the parameter UM/ν tends to a limiting form, as $UM/\nu \to \infty$, for values of κ within the quasi-equilibrium range but outside the equilibrium range; this has already been inferred from measurements of the correlation function when plotted as in fig. 7.8.

The reduction to the conditions of the equilibrium range when κ is very large can be understood from the dynamical equation (5.5.11) for the spectrum function. In view of (7.3.5) this equation becomes

$$\left(\frac{1}{\lambda}\frac{d\lambda}{dt} + \frac{1}{u^2}\frac{du^2}{dt}\right)E_{qe} + \kappa\frac{d\lambda}{dt}\frac{\partial E_{qe}}{\partial z} = \frac{1}{u^2\lambda}T(\kappa, t) - 2\nu\kappa^2 E_{qe},$$

that is

$$-\tfrac{1}{2}E_{qe} + \tfrac{1}{2}z\frac{\partial E_{qe}}{\partial z} = \frac{t}{u^2\lambda}T(\kappa, t) - \tfrac{1}{8}z^2 E_{qe}, \tag{7.3.8}$$

where $z = \lambda\kappa$. When z is very large, the term $z^2 E_{qe}$ will always be large compared with the first term on the left side and will, for most functional forms of E_{qe}, be large compared with the second term on the left. Only in cases in which E_{qe} decreases with z as rapidly as $e^{-\alpha z^2}$ (α constant) will the terms on the left side fail to be negligible. We cannot exclude such cases rigorously, but the effect of the non-linear transfer of energy is to spread the energy as widely as possible, and they may be rejected provisionally as being incompatible with finite positive values of $T(\kappa, t)$. Then the left side of (7.3.8) may be neglected, giving the dynamical equation appropriate to the equilibrium range. More precise information about the location of the boundary between the equilibrium and quasi-equilibrium ranges cannot be obtained without making assumptions about the functional form of the transfer term.

When κ is very small, we know from the exact analysis of § 5.3 that the transfer term $T(\kappa, t)$ and the viscous term $-2\nu\kappa^2 E(\kappa, t)$ are both of smaller order than the term $\partial E(\kappa, t)/\partial t$ in the dynamical equation (6.6.1). Consequently, if the similarity relations of the quasi-equilibrium were to extend to very small values of κ (a hypothetical case only), only the terms on the left side of (7.3.8) would survive and the required form of the spectrum there would be

$$E_{qe} \propto \lambda\kappa, \tag{7.3.9}$$

that is, from (7.3.6),
$$E(\kappa, t) \propto u^2\lambda^2\kappa, \tag{7.3.10}$$

which is independent of t.† We found in § 5.3 that the continuity condition required $E(\kappa, t)$ to vary as κ^4 near $\kappa = 0$, so that quite apart from the inability of the very big eddies to take part in a quasi-equilibrium based on inertial exchange of energy, their distribution of energy is necessarily different from that required by a quasi-equilibrium.

7.4. The equilibrium at large wave-numbers for moderate Reynolds numbers

The universal equilibrium theory described in Chapter VI was arrived at from a consideration of the nature of turbulent motion at very large Reynolds number. When the Reynolds number is so large that the energy-containing range of wave-numbers and the range in which the viscous dissipation occurs are widely separated, it was found possible to postulate that the motion associated with large wave-numbers has a universal statistical form determined uniquely by the parameters ν and ϵ. This possibility rests on the notion that the motions associated with widely separated wave-numbers are statistically independent, so that the only connection between the equilibrium range and the remainder of the turbulence lies in the inertial transfer of energy across the spectrum at a rate ϵ.

† The transformation of the E against κ graph represented by (7.3.5) is equivalent to a stretching of both ordinate and abscissa by a factor proportional to $t^{\frac{1}{2}}$ (as in fig. 7.6); if this transformation is to make the spectrum curves at different values of t coincident, the spectrum curves are necessarily linear near the origin, with the same slope.

When the Reynolds number is not very large, so that the energy-containing range and the dissipation range are not widely separated, we cannot postulate that these two ranges are statistically independent and a universal equilibrium is not to be expected. Nevertheless, if it is true that Fourier components with widely separated wave-numbers tend to be statistically independent even when one of these wave-numbers lies beyond the dissipation range, there will always be *some* large wave-numbers for which the associated motion is statistically steady and independent of the mechanical agencies which generated the turbulence. Provided the inertia forces are strong enough to transfer energy to wave-numbers $\kappa \gg \kappa_0$, some kind of equilibrium theory should apply to these large wave-numbers, irrespective of whether all the viscous dissipation occurs in this range. The difficulty about a theory appropriate to these moderate Reynolds numbers is that we do not know what parameters determine the motion associated with this new equilibrium range of wave-numbers. For large Reynolds numbers there is a clear guide to the choice of parameters in the fact that the rate of transfer of energy into the equilibrium range is ϵ. It might perhaps be speculated that the determining parameter at moderate Reynolds number is the contribution to the rate of dissipation from Fourier components with wave-numbers lying within the equilibrium range. However, this begs the question a little, since we have no means of predicting what fraction of the total dissipation occurs in the equilibrium range, and it seems an unsuitable hypothesis in view of the fact that there is no sharp boundary dividing the equilibrium range from smaller wave-numbers; moreover, the speculation receives no support from measurements. It seems unlikely that any theory which considers the equilibrium range, as such, in isolation from the remainder of the motion is likely to be successful at moderate Reynolds numbers.

However, we can obtain some help from the observations of the energy-containing eddies in the turbulence generated by grids, described earlier in this chapter. It was shown there that, apart from a range of small wave-numbers containing about 25% of the energy, the turbulence maintains a statistical similarity described by the three factors ν, ϵ and t during decay. Moreover, this similarity was found to exist over a wide range of Reynolds numbers *lower* than that necessary for the existence of a universal equilibrium. The absence of any severe restriction on the Reynolds number in the case of the quasi-equilibrium is not altogether unexpected, because it is not an essential part of the ideas set out in the preceding section that all the dissipation should occur in the equilibrium range. Only a small extension of the ideas is required to lead us to the hypothesis of a quasi-equilibrium at moderate Reynolds numbers (we should probably need to exclude Reynolds numbers so low that some of the dissipation occurs outside the quasi-equilibrium range), as Goldstein (1951) has emphasized. Consequently if it is true that a quasi-equilibrium always merges into an equilibrium as $\kappa \to \infty$, the parameter t becoming less and less relevant—some relevant evidence is presented below—we have the conclusion, three parts empirical, one part analytical, that an equilibrium determined by the parameters ν and ϵ exists at moderate Reynolds numbers. Presumably the equilibrium which exists at moderate Reynolds numbers will be identical with part (at the high wave-number end) of the universal equilibrium which exists at large Reynolds numbers.

Not many measurements of the spectrum function at large values of κ in turbulence generated by grids have been made, so that the above ideas must remain tentative. The only available measurements seem to be those of Stewart and Townsend (1951), made at Reynolds numbers $UM/\nu = 2625$, 5250 and 10,500, and already described partially in fig. 7.7. The different (one-dimensional) spectrum curves at different Reynolds numbers in fig. 7.7 seem to be coming together as $\kappa \to \infty$, but the spectrum function has small values in this region and small differences between the curves could be hidden. Stewart and Townsend therefore measured the one-dimensional spectra of $\partial u_1/\partial x_1$, $\partial^2 u_1/\partial x_1^2$ and $\partial^3 u_1/\partial x_1^3$—effectively giving measurements of the functions $\kappa^2\phi(\kappa, t)$, $\kappa^4\phi(\kappa, t)$ and $\kappa^6\phi(\kappa, t)$—and compared them at the above three Reynolds numbers. It was found that there were detectable variations in the functions $\kappa^2\phi(\kappa, t)$ and $\kappa^4\phi(\kappa, t)$ at the different Reynolds numbers when plotted in the manner of fig. 7.7, but not in $\kappa^6\phi(\kappa, t)$, the measurements of which are reproduced in fig. 7.9.† The value of

† An interesting observation was that the measured values of $\kappa^6\phi(\kappa, t)$ at distances of $30M$ and $40M$ from the grid, for all Reynolds numbers, fell below

κ at which the spectra at these different Reynolds numbers can be said to coincide is not well defined, but the figures make it clear that the coincidence improves as $\kappa \to \infty$. Hence they support the hypothesis that for turbulence generated by a grid at moderate Reynolds numbers the motion associated with a range of sufficiently large wave-numbers has a universal statistical form which is determined

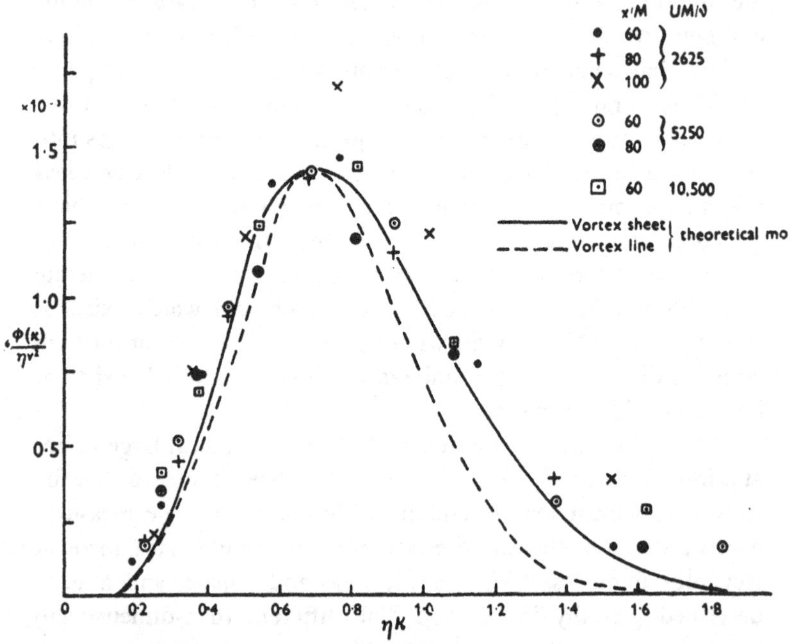

Fig. 7.9. Spectrum function (one-dimensional), at high wave-numbers for different Reynolds numbers (after Townsend, 1951*b*).

by ν and ϵ and which is presumably identical† with part of the equilibrium which exists at large Reynolds numbers.

It should be noted that there are as yet no grounds for supposing that the motion associated with the equilibrium range has the same

the curve shown in fig. 7.9. The interpretation is that at the stage represented by $x = 40M$ the decay was not quite far enough advanced for the inertial transfer of energy to have built up the energy at these large wave-numbers to its equilibrium level.

† Which is the reason why it is permissible to use measurements at moderate Reynolds numbers to check theoretical predictions about $E(\kappa, t)$ at the larger values of κ within the universal equilibrium range, as was done in § 6.6 in connexion with Heisenberg's postulated form of the energy transfer function.

statistical form for all types of turbulent motion at moderate Reynolds number. The dependence of the equilibrium range on ν and ϵ at moderate Reynolds number, as described above, is essentially a consequence of the existence of a quasi-equilibrium at lower wave-numbers, which is probably a circumstance peculiar to decaying homogeneous turbulence. In the case of, say, statistically steady turbulent flow through a pipe it is very probable that the energy-containing eddies are all influenced by the boundary conditions of the motion, so that at moderate Reynolds numbers for which the energy-containing range and dissipation range are not widely separated the dissipation ϵ does not occur wholly within a range which is characterized by the same statistical conditions as in the case of grid turbulence. Under these conditions the parameter ϵ is unlikely to have the same significance in the two cases. The equilibrium at moderate Reynolds numbers is probably 'universal' for grid turbulence alone, unlike the equilibrium at large Reynolds numbers.

A simple model of that part of the turbulent motion represented by Fourier components with wave-numbers beyond the dissipation range has recently been proposed by A. A. Townsend (1951*b*), and since it is, in effect, a postulate about the equilibrium which exists at these very large wave-numbers at all Reynolds numbers it can appropriately be described here. Townsend's idea is that the mechanical equilibrium between inertial transfer and viscous dissipation, at wave-numbers $\kappa \gg \kappa_d$, is a reflexion of a balance between the tendency for the straining motion (which arises from wave-numbers of the order of κ_d) to extend and concentrate vortex sheets or lines and the tendency for viscosity to diffuse and weaken them. On this basis he calculates the form of $E(\kappa)$ for $\kappa \gg \kappa_d$ from a random distribution of these elementary, stationary, small-scale velocity fields. Consider first a part of the fluid where the local relative motion consists instantaneously of principal rates of strain α, $-\alpha$, 0 $(\alpha > 0)$ in the directions of the x_1-, x_2- and x_3-axes respectively. Then any small perturbation (for which quadratic terms can be neglected) to this vorticity distribution will tend to be distorted by the plane straining motion into a vortex sheet in the x_1-, x_3-plane, with the perturbation vorticity in the direction of the x_1-axis. If the basic straining motion remains approximately steady for long enough (the necessary time being smaller for higher wave-number components of the perturbation), the velocity distribution of the perturbation tends asymptotically to

$$\frac{du_3}{dx_2} = \omega_0 \exp\left[-\frac{\alpha x_2^2}{2\nu}\right], \quad u_1 = u_2 = 0, \tag{7.4.1}$$

representing a vortex sheet of finite thickness, where ω_0 is a measure of the strength of the vorticity perturbation. Similarly, a basic straining motion with two positive principal rates of strain converts (asymptotically) a vorticity perturbation into two superimposed vortex sheets, and one with two negative principal rates of strain converts a perturbation into a vortex line.

We now imagine a large number of such asymptotic vorticity perturbations to be distributed randomly (with respect to both position and orientation) throughout the fluid, with the approximation that the basic straining motion has the same intensity everywhere. It is then a simple matter to calculate the spectrum function describing the perturbations; for the above case of a plane straining motion the one-dimensional spectrum function is found to have the same form at large values of κ as

$$\phi(\kappa) = \frac{A}{\kappa^2}\int_0^1 (1-m^2)\exp\left[-\frac{\kappa^2\nu}{\alpha m^2}\right] dm, \tag{7.4.2}$$

where A is not determined by the model. If the value of α is taken as $(\epsilon/4\nu)^{\frac{1}{2}}$, so that the rate at which the energy of the straining motion is being dissipated by viscosity is ϵ, the function $\kappa^6\phi(\kappa)$ as given by (7.4.2) has the shape shown in fig. 7.9. Similar calculations for a straining motion which is symmetrical about an axis lead to the other curve shown in fig. 7.9; the fact that the difference is so slight suggests that the shape for an arbitrary straining motion would not be very different. Both of these curves in fig. 7.9 involve a single unknown parameter, viz. a scale factor of the ordinate, like A in (7.4.2), which has been chosen to place the maxima of the curves at about the same value of $\kappa^6\phi(\kappa)$ as for the experimental points. The value of κ/κ_d at which the maxima occur involves no disposable parameters, and the agreement with experiment therefore seems to be remarkably good. However, it should be kept in mind that the assumption that the straining motion is spatially uniform does not permit us to use the model to get information about wave-numbers in the dissipation range, and since the experimental variation of $\kappa^2\phi(\kappa)$ with κ shows a maximum at about $\kappa = 0{\cdot}2\kappa_d$, it is only for wave-numbers large compared with this value that the model is relevant. The model also receives some support from the observed spottiness of the spatial distribution of high-order derivatives of the velocity (see § 8.4).

The value of the model is that it supplies a plausible picture of the equilibrium at very large wave-numbers, and in particular of the way in which energy is transferred to these large wave-numbers. The mechanism of the transfer of energy is here quite different from the eddy friction mechanism proposed by Heisenberg; mathematically this difference leads to the transfer of energy to very large wave-numbers being proportional to the root-mean-square of the total vorticity with Townsend's model, but to the mean-square with Heisenberg's expression for the transfer. It should also be noticed that according to the model the Fourier components with very large wave-numbers κ ($\gg \kappa_d$) are not statistically independent of the Fourier components with wave-numbers of order κ_d from which they receive energy, since the straining motion produces a definite alinement of the vorticity perturbation; this is not inconsistent with the 'disorder' hypothesis of § 6.3, since for the latter to apply at least one of the two Fourier components concerned must be dominated by inertia forces. On the other hand, Heisenberg's transfer expression assumes implicitly (in virtue of the representation of the transfer as a friction process) that the pair of Fourier components involved in any exchange of energy are statistically independent; since the Fourier components in the range in which $E(\kappa)$ is found to vary as κ^{-7} receive their energy from wave-numbers of order κ_d, the assumption of a friction process is here going beyond the disorder hypothesis as we have formulated it.

7.5. Heisenberg's form of the energy spectrum in the quasi-equilibrium range

The intuitive hypothesis about the form of the transfer term in the dynamical equation for the spectrum which v. Weizsäcker and Heisenberg put forward has already been described in § 6.6. Heisenberg originally used this hypothesis to deduce the form of the energy spectrum in the equilibrium range of wave-numbers.

When it was realized later that the experimental results supported the idea of a stable statistical state of some, at least, of the energy-containing eddies during decay, Heisenberg (1948*b*) used the same hypothesis about the transfer term in order to determine the energy spectrum in the quasi-equilibrium range of wave-numbers. We saw in § 6.6 that the computed spectrum in the equilibrium range was not in agreement with experiments, so far as the latter could be made under the relevant conditions, but, somewhat surprisingly, the agreement in the quasi-equilibrium range has been found by I. Proudman (1951) to be reasonably good, and a brief description of the calculation of the spectrum will therefore be given here.

The dynamical equation for the spectrum when the turbulence is isotropic can be written (see (6.6.3) and (6.6.4)) as

$$\frac{\partial}{\partial t}\int_\kappa^\infty E(\kappa'',t)\,d\kappa''=S(\kappa,t)-2\nu\int_\kappa^\infty \kappa''^2E(\kappa'')\,d\kappa''. \tag{7.5.1}$$

We are postulating the existence of a quasi-equilibrium for values of κ which are not too small, in which case all the terms in (7.5.1) are functions of κ, ϵ, ν and t only. As already shown in § 7.3, this implies that $E(\kappa, t)$ has any one of the equivalent forms (7.3.4), (7.3.5) or (7.3.6); we use the form (7.3.5), viz.

$$E(\kappa,t)=u^2\lambda E_{qe}\left(\lambda\kappa,\frac{u\lambda}{\nu}\right),$$

where $\lambda^2=10\nu t$ and $u\lambda/\nu$ is independent of t; the spectrum function can therefore be written as

$$E(\kappa,t)=\left(\frac{\nu^3}{t}\right)^{\frac{1}{2}}F\left(\chi,\frac{u\lambda}{\nu}\right), \tag{7.5.2†}$$

† J. Rotta (1950) and N. R. Sen (1951) have investigated the consequences, using Heisenberg's form of the transfer term, of assuming different kinds of similarity (which of necessity cannot embrace values of κ in the dissipation range, except in the case (7.5.2)) during decay, that is, of reducing $E(\kappa, t)$ to a single independent variable involving κ and t by transformations different from (7.5.2). A. N. Kolmogoroff (1941*b*) proposed one such similarity law (which is chosen so as to make the range of similarity include both the energy-containing eddies and the largest eddies for which $E(\kappa)\approx C\kappa^4$; the mechanical basis for such a similarity is not clear in view of the inability of Fourier components with very small values of κ to interact with other Fourier components), which has been discussed further by F. N. Frenkiel (1948*b*) and T. von Kármán (1948*b*). All these similarity transformations invoke energy decay laws different from the observed law (7.1.1) (that suggested by Kolmogoroff requiring $u^2\propto t^{-\frac{10}{7}}$). In general they are inconsistent with the view, which has been put forward here, that the similarity of the energy-containing eddies is closely linked with the even simpler conditions which exist at larger wave-numbers in the equilibrium range.

where $\chi = \kappa(\nu t)^{\frac{1}{2}}$. Similarly, the transfer function $S(\kappa, t)$ can be written as

$$S(\kappa, t) = \frac{\nu}{t^2} S_{qe}\left(\chi, \frac{u\lambda}{\nu}\right) \tag{7.5.3}$$

for values of κ such that $S(\kappa, t)$ is determined by the motion associated with the quasi-equilibrium range, so that (7.5.1) becomes

$$-\tfrac{1}{2}\chi F(\chi) - \int_{\chi}^{\infty} F(\chi'')\,d\chi'' = S_{qe}(\chi) - \int_{\chi}^{\infty} 2\chi''^2 F(\chi'')\,d\chi''. \tag{7.5.4}$$

The parameter $u\lambda/\nu$ is constant during the decay and depends on the initial conditions (viz. on UM/ν for grids of the same geometrical shape) only, so that (7.5.4) is an ordinary integro-differential equation in χ which should yield one solution for each value of $u\lambda/\nu$.

Equation (7.5.4) is to be solved subject to the condition

$$\int_0^{\infty} 2\chi^2 F(\chi)\,d\chi = t^2 \int_0^{\infty} 2\kappa^2 E(\kappa)\,d\kappa$$

$$= \frac{\epsilon t^2}{\nu} = 0{\cdot}15\left(\frac{u\lambda}{\nu}\right)^2 = R, \tag{7.5.5}$$

where R is constant during decay and has been found, for turbulence generated by square-mesh grids, to have the value $\frac{1}{90}\frac{UM}{\nu}$.
Equation (7.5.4) can be regarded as formally valid for all values of χ, on the understanding that the solution at small values of χ (i.e. of κ) describes the hypothetical spectrum which would be required by the existence of a quasi-equilibrium covering *all* the energy of the turbulence. Then letting $\chi \to 0$ in (7.5.4) we find the integral condition

$$\int_0^{\infty} F(\chi)\,d\chi = \int_0^{\infty} 2\chi^2 F(\chi)\,d\chi = R, \tag{7.5.6}$$

which is a consequence of the similarity transformation (7.5.2). Near $\chi = 0$ the solution is

$$F(\chi) \propto \chi, \tag{7.5.7}$$

as found earlier for an arbitrary form of the transfer term.

The additional hypothesis made by Heisenberg is that $S_{qe}(\chi)$ has the form

$$S_{qe}(\chi)=\gamma\int_{\chi}^{\infty}\sqrt{\frac{F(\chi')}{\chi'^3}}\,d\chi'\int_{0}^{\chi}2\chi''^2F(\chi'')\,d\chi'', \qquad (7.5.8)$$

as originally proposed by v. Weizsäcker for the equilibrium range; the arguments for and against the expression (7.5.8) were described in §6.6 and apply equally well to its use in the quasi-equilibrium range. The equation to be solved is then

$$-\tfrac{1}{2}\chi F(\chi)-\int_{\chi}^{\infty}F(\chi'')\,d\chi''=\gamma\int_{\chi}^{\infty}\sqrt{\frac{F(\chi')}{\chi'^3}}\,d\chi'\int_{0}^{\chi}2\chi''^2F(\chi'')\,d\chi'' -\int_{\chi}^{\infty}2\chi''^2F(\chi'')\,d\chi''. \qquad (7.5.9)$$

If the terms on the left side of (7.5.9) are neglected, giving the equation applicable to the equilibrium range, the solution has the form (6.6.15); this solution is asymptotic to a power law ($F(\chi)\sim\chi^{-7}$) as $\chi\to\infty$, so that for Heisenberg's transfer term it is certainly true that at sufficiently large values of κ the terms on the right side of (7.5.9) dominate those on the left—in other words, that the quasi-equilibrium reduces to an equilibrium—for an arbitrary value of the Reynolds number. Hence provided that there is in fact only one solution of (7.5.9) for each value of the Reynolds number R, such that $F(0)=0$, all the solutions of (7.5.9) are linear in χ near $\chi=0$ and vary as χ^{-7} as $\chi\to\infty$, while for solutions for very large values of R there will be a range of (large) values of χ for which $F(\chi)\propto\chi^{-\frac{5}{3}}$.

An analytical solution of (7.5.9) has not been given, but S. Chandrasekhar (1951) has obtained the solution numerically† for several values of R. By using y and $g(y)$ as independent and dependent variables respectively, where

$$y=\gamma^2\int_{0}^{\chi}\chi''^2F(\chi'')\,d\chi'', \quad g=\gamma^2\chi^3F(\chi), \qquad (7.5.10)$$

it is possible to convert (7.5.9) into the differential equation

$$g^{\frac{3}{2}}g''+2y(4+g')+2g^{\frac{1}{2}}(4-g')-8g=0, \qquad (7.5.11)$$

† J. Rotta (1949) claims to have obtained a numerical solution for the case of infinite Reynolds number and he has published a curve showing the consequent form of the one-dimensional spectrum function $\phi(\kappa)$, although no details of the calculation are given. This work was independent of Heisenberg's paper (1948*b*).

where dashes denote differentiation with respect to y. The boundary conditions for this equation are obtained from (7.5.6) and (7.5.7) above as

$$\left.\begin{aligned} g&=0 \quad \text{when} \quad y=\tfrac{1}{2}\gamma^2 R,\\ g&\to 4y \quad \text{as} \quad y\to 0. \end{aligned}\right\} \tag{7.5.12}$$

Near $y=0$ the second approximation is found from (7.5.11) to be

$$g(y)=4y+y^{\frac{4}{3}}(a+\tfrac{4}{3}\log y), \tag{7.5.13}$$

where a is arbitrary, and the numerical solution is obtained by a forward integration from this point. The solution is determined completely when a value of a is chosen, and the corresponding value of $\gamma^2 R$ is found from the first of the boundary conditions (7.5.12).

Three of the solutions† obtained by Chandrasekhar‡ (1949a), and one obtained by Proudman (1951) (for $\gamma^2 R=21\cdot 9$), by the same method are shown in fig. 7.10 in the form of $\gamma^2\alpha^{-\frac{3}{2}}F(\chi)$ as a function of $\alpha^{\frac{1}{2}}\chi$ for several values of $\gamma^2 R$. α is the constant of integration which arises in the determination of $F(\chi)$ in terms of g and y by means of the relations (7.5.10), and has been chosen in such a way that the solutions have length scales which do not vary greatly with R. A convenient way of doing this is to choose α so that all the curves in fig. 7.10 have the same slope (viz. 4) at the origin (which implies that the area under each curve in the figure is $\gamma^2 R/\alpha$); these curves thus differ from the solutions of (7.5.9) by a linear transformation. One of the solutions obtained by Chandrasekhar is for the limiting case $\gamma^2 R\to\infty$ and, as shown in the figure, it is asymptotic, as $\kappa\to\infty$, to the equilibrium spectrum

$$F(\chi)\sim\left(\frac{8R}{9\gamma}\right)^{\frac{2}{3}}\chi^{-\frac{5}{3}}, \tag{7.5.14}$$

i.e.

$$\gamma^2\alpha^{-\frac{3}{2}}F(\chi)\sim\left(\frac{8\gamma^2 R}{9\alpha}\right)^{\frac{2}{3}}(\chi\sqrt{\alpha})^{-\frac{5}{3}},$$

† The solution is not absolutely complete until a value of γ is given, but since γ occurs essentially in the combination $\gamma^2 R$ and so determines only the Reynolds number of any one of the solutions in fig. 7.10, it is sufficient for most purposes to know that γ will be of order unity.

‡ Note that Chandrasekhar's Reynolds number R is not the same as that used here. The method of making use of solutions of (7.5.11) which is described herein was devised by Proudman.

where the value of α appropriate to $\gamma^2 R = \infty$ was found by evaluation of the area under the curve representing the solution for this case.

The comparison between these computed solutions and the measured statistical functions has been undertaken by Proudman (1951). It is clear from fig. 7.10 that, as noticed by Stewart from measurements (see § 7.3), the part of the spectrum containing most

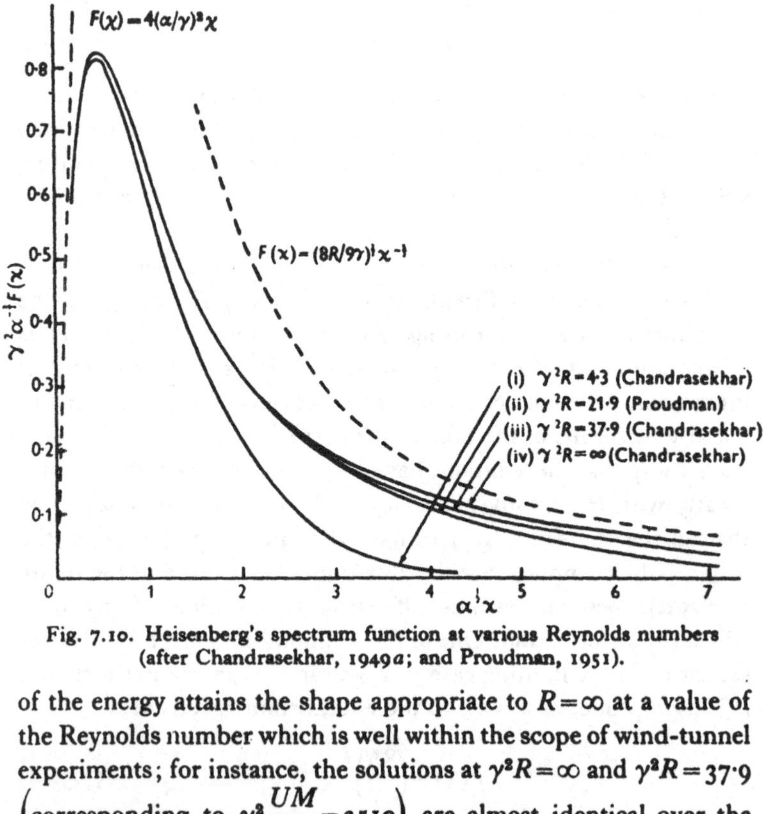

Fig. 7.10. Heisenberg's spectrum function at various Reynolds numbers (after Chandrasekhar, 1949*a*; and Proudman, 1951).

of the energy attains the shape appropriate to $R = \infty$ at a value of the Reynolds number which is well within the scope of wind-tunnel experiments; for instance, the solutions at $\gamma^2 R = \infty$ and $\gamma^2 R = 37{\cdot}9$ $\left(\text{corresponding to } \gamma^2 \dfrac{UM}{\nu} = 3510\right)$ are almost identical over the energy-containing range of wave-numbers. Consequently, the theoretical solution for $\gamma^2 R = \infty$ may be compared with measurements at any sufficiently high Reynolds number. The two Reynolds numbers chosen for the comparison were $\gamma^2 R = \infty$ and $\gamma^2 R = 21{\cdot}9$, and Proudman calculated the velocity correlation functions—which can be measured more accurately, in the energy-containing range, than the spectrum—from the two appropriate curves in fig. 7.10.

The data for the comparison were taken from results for square-mesh grids published by Townsend and by Stewart. The comparison between theory and experiment is complicated, chiefly owing to the presence of the unknown constant γ in the theoretical solution, and it is not possible to reproduce the details here. Proudman's conclusion is that the theory predicts satisfactorily the general shape of the correlation functions (excluding large values of r, which lie outside the quasi-equilibrium range) at the two Reynolds numbers, provided γ is chosen as $0{\cdot}45 \mp 0{\cdot}05$. As an example of the agreement obtained, fig. 7.11 shows the comparison between the calculated longitudinal correlation function $f(r)$, for

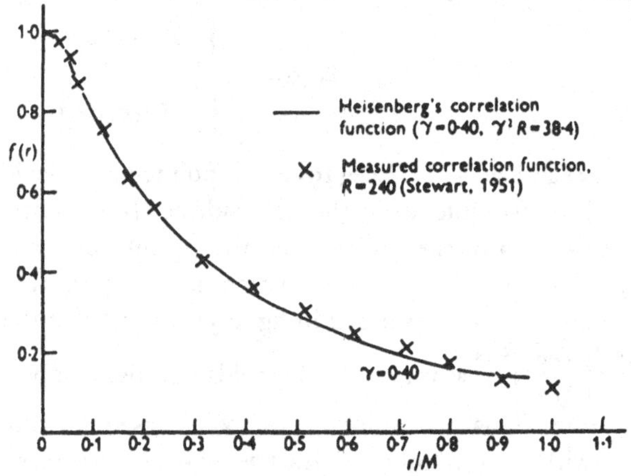

Fig. 7.11. Comparison between theoretical and measured correlation functions (after Proudman, 1951).

$\gamma^2 R = 38{\cdot}4$ and $\gamma = 0{\cdot}40$ (obtained from the theoretical curves at other values of $\gamma^2 R$ by the process of adjustment described at the end of §7.2), and some measurements of this function made at $R = 240$.

Although the agreement with experiment is very promising, it cannot yet be regarded as sufficiently decisive to permit the inference that Heisenberg's expression for the transfer is essentially accurate. The general shape of correlation functions is notoriously insensitive to theoretical assumptions, and the agreement has been assisted to some extent by the fact that γ is a disposable constant. What we can say with confidence is that the measurements of the correlation functions in the energy-containing range are not inconsistent with v. Weizsäcker's and Heisenberg's hypothesis about the energy transfer, provided $\gamma = 0{\cdot}45 \mp 0{\cdot}05$. At very large values of κ, beyond the dissipation range of wave-numbers, there is, as we noted in § 6.6, disagreement between the measured spectrum and the asymptotic form of Heisenberg's spectrum function. There are other indications of this lack of validity at large wave-numbers. The form of the vorticity equation (5.5.8) appropriate to large Reynolds numbers at which a universal equilibrium exists is

$$\frac{\overline{\left(\frac{\partial u_1}{\partial x_1}\right)^3}}{\left[\overline{\left(\frac{\partial u_1}{\partial x_1}\right)^2}\right]^{\frac{3}{2}}} = \frac{k_0'''}{(-f_0'')^{\frac{3}{2}}} = -\frac{2\nu}{u}\frac{f_0^{\text{iv}}}{(-f_0'')^{\frac{3}{2}}}$$

$$= -\tfrac{3}{7}(30)^{\frac{1}{2}}\,\nu\frac{\int_0^\infty \kappa^4 E(\kappa)\,\mathrm{d}\kappa}{\left[\int_0^\infty \kappa^2 E(\kappa)\,\mathrm{d}\kappa\right]^{\frac{3}{2}}} \qquad (7.5.15)$$

from (3.4.25) and, as T. D. Lee (1950) has pointed out, the left side of (7.5.15) is measurable, while the right side can be evaluated from Heisenberg's form of the spectrum in the equilibrium range (see (6.6.15)). The result of the calculation is $-1{\cdot}52\gamma$, while the measurements (already described in fig. 6.3) suggest that the value of $\overline{\left(\frac{\partial u_1}{\partial x_1}\right)^3} \Big/ \left[\overline{\left(\frac{\partial u_1}{\partial x_1}\right)^2}\right]^{\frac{3}{2}}$ at very large Reynolds number is about $-0{\cdot}3$, giving $\gamma = 0{\cdot}2$. This determination of γ is based on quantities associated with wave-numbers at least as large as those responsible for the dissipation, so that Heisenberg's hypothesis leads to predictions about the energy-containing and dissipation ranges of wave-numbers which can be valid only if substantially different values of γ for the two ranges are used. Such a possibility is not worth admitting (since it would undermine the physical basis of the hypothesis about the energy transfer), and we must conclude that Heisenberg's spectrum may represent the facts accurately in the energy-containing range but cannot do so at larger wave-numbers in the equilibrium range. There may be some significance in the possibility of agreement with experiment only under conditions such that viscous forces are negligible; the ideas underlying the hypothesis are concerned entirely with inertial effects, and it may be that the absence of viscous forces is a necessary condition for their validity.

CHAPTER VIII

THE PROBABILITY DISTRIBUTION OF **u**(**x**)

8.1. The experimental evidence

In the two preceding chapters the discussion has inevitably centred around the physically important spectrum tensor. This is the statistical quantity which lends itself most readily to the formulation of hypotheses about the mechanism of decay of the turbulence. However, there are some other statistical quantities whose meanings are readily understood and which are relevant to the basic mechanical processes of the motion. Almost all of these quantities are determined by the probability distribution of the velocity or its derivatives at one point, or by the joint-probability distribution of the velocity at two points. The intention in this chapter is to consider these two probability distributions in the hope that some light will be thrown on the form of the all-embracing joint-probability distribution of the velocities at any number of points. This latter distribution will be referred to simply as the distribution of the function **u**(**x**).

We begin with the probability density function of the velocity at an arbitrary point and at a given time of decay. A. A. Townsend (1947) has devised a method of measuring this function directly by means of a suitable electrical analysis of the output of a hot-wire anemometer placed at the point in question. A sample of the results obtained for the velocity component u_1 in the direction of the stream for the turbulence generated by a square-mesh grid is shown in fig. 8.1. These measurements are fitted very closely by the normal or Gaussian probability density function shown in the figure, and the same is true of measurements of u_1, u_2 or u_3 at different times of decay and at different Reynolds numbers. A sensitive test of the closeness of the fit over the outer parts of the curve is provided by measurements of the factor $\overline{u_1^4}/(\overline{u_1^2})^2$, which measures the relative flatness of the distribution. Townsend has found that this factor lies between 2·9 and 3·0, within the limits of experimental error, as compared with the value 3·0 appropriate to a normal distribution of **u**(**x**). Measurements of the factor $\overline{u_1^3}/(\overline{u_1^2})^{\frac{3}{2}}$, which measures the relative skewness of the distribution, gave the value zero. The existence of an approximately normal distribution of the velocity at one point has been known for many years (for example, see Simmons and Salter, 1938), and was one of the first experimental results concerning turbulent motion to be established.

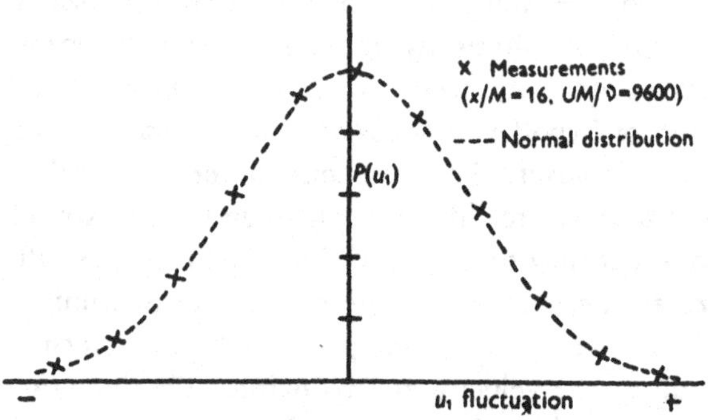

Fig. 8.1. Probability density function of u_1.

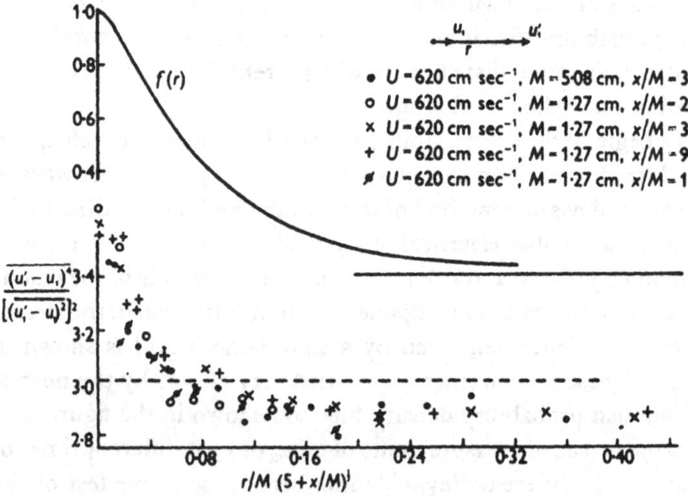

Fig. 8.2. Flatness factor of the distribution of $u_1' - u_1$.

The same amount of information about the joint-probability distribution of the velocities at two different points is not yet available, but enough measurements have been made to show that in general it is not accurately normal.† Fig. 8.2 shows measurements

† For the definition of a normal joint-probability distribution, see, for example, H. Cramér, op. cit. (p. 98).

(made by R. W. Stewart) of the factor

$$\overline{(u_1'-u_1)^4}/[\overline{(u_1'-u_1)^2}]^2, \qquad (8.1.1)$$

for different values of r, where

$$u_1 \equiv u_1(x_1, x_2, x_3) \quad \text{and} \quad u_1' = u_1(x_1+r, x_2, x_3),$$

the x_1-axis being directed downstream, and at different stages of the decay behind a grid of square mesh. The length $M\left(\frac{x}{M}+5\right)^{\frac{1}{2}}$, where x is the distance from the grid, is used as the unit of r, because the measured longitudinal velocity correlation $f(r) = \overline{u_1'u_1}/\overline{u_1^2}$, which is shown for comparison, then has approximately the same shape at all the relevant stages of the decay. If the joint-probability distribution of u_1 and u_1' were accurately normal, the flatness factor of $u_1'-u_1$ would be 3·0. The measured flatness factors have approximately this value for r greater than about $0{\cdot}05M\left(\frac{x}{M}+5\right)^{\frac{1}{2}}$, at which value of r the correlation coefficient $f(r)$ is about 0·6. At values of r which are so large that u_1 and u_1' are statistically independent, the flatness factor (8.1.1) reduces to $1{\cdot}5+\frac{1}{2}\frac{\overline{u_1^4}}{(\overline{u_1^2})^2}$, and the measurements in this range merely recover the result noted in the previous paragraph. At small values of r the measured flatness factor departs significantly from 3·0 and has a maximum value of about 3·6 at $r=0$.

Results which are qualitatively similar have been found for the skewness factor of the probability distribution of $u_1'-u_1$, i.e. for

$$\overline{(u_1'-u_1)^3}/[\overline{(u_1'-u_1)^2}]^{\frac{3}{2}}. \qquad (8.1.2)$$

R. W. Stewart (1951) has published measurements of this quantity for different values of r and for different values of the Reynolds number UM/ν of the grid, which are shown in fig. 8.3. The values of the skewness depend significantly on Reynolds number, but for each Reynolds number the trend is from an extreme value at $r=0$ to the value appropriate to a normal distribution (viz. zero) at large values of r. Although the skewness falls steeply at small values of r it does not reach the value appropriate to a normal distribution at values of r at which $f(r)$ is still well above zero (unlike the flatness factors shown in fig. 8.2).

When r is sufficiently small, the probability distribution of $(u_1' - u_1)$ is identical with that of $\partial u_1/\partial x_1$. (The points at $r = 0$ in figs. 8.2 and 8.3 were in fact obtained from an (electrical) differentiation of the velocity recorded at a point fixed relative to the grid. The agreement with the remainder of the curve provides further support for the fundamental assumption that the velocity is a continuous function of position and for the approximation that the velocity variations at the point fixed relative to the grid are the same

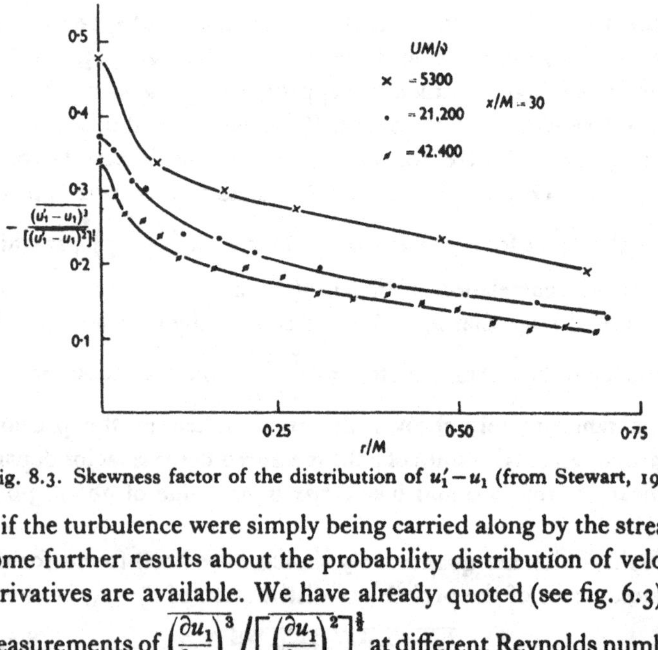

Fig. 8.3. Skewness factor of the distribution of $u_1' - u_1$ (from Stewart, 1951).

as if the turbulence were simply being carried along by the stream.) Some further results about the probability distribution of velocity derivatives are available. We have already quoted (see fig. 6.3) the measurements of $\overline{\left(\frac{\partial u_1}{\partial x_1}\right)^3} \Big/ \left[\overline{\left(\frac{\partial u_1}{\partial x_1}\right)^2}\right]^{\frac{3}{2}}$ at different Reynolds numbers, which suggest that the limiting value for very large Reynolds numbers is about $-0{\cdot}3$. Fig. 8.4 shows a sample complete distribution of $\partial u_1/\partial x_1$ for turbulence generated by a square-mesh grid, obtained by A. A. Townsend by an electrical analysis of the differentiated signal from the hot-wire anemometer. The departures from a normal curve with the same standard deviation are consistent with the measurements of the skewness and flatness factors mentioned above. Small positive values of $\partial u_1/\partial x_1$ are more probable than small negative values, but this positive contribution to the skewness is more than balanced by the higher frequency of occurrence of negative, as compared with positive, large values of $\partial u_1/\partial x_1$. Townsend has also measured $\overline{\left(\frac{\partial u_2}{\partial x_1}\right)^4} / \left[\overline{\left(\frac{\partial u_2}{\partial x_1}\right)^2}\right]^2$ (the direction of the x_2-axis being at right angles to the x_1-axis but otherwise arbitrary), and finds its value to be about 3·3, with no marked variation with time of decay or Reynolds number.

A further set of experimental results concerning the probability distribution of $\mathbf{u}$ and its derivatives at one point is provided by some measurements of

$$\overline{\left(\frac{\partial^n u_1}{\partial x_1^n}\right)^4} \Big/ \left[\overline{\left(\frac{\partial^n u_1}{\partial x_1^n}\right)^2}\right]^2 \tag{8.1.3}$$

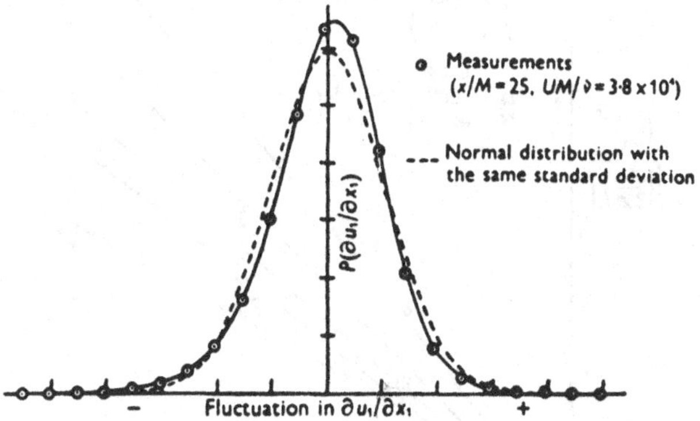

Fig. 8.4. Probability density function of $\partial u_1/\partial x_1$.

at different Reynolds numbers for values of n up to 3 (Batchelor and Townsend, 1949), which are reproduced in fig. 8.5. Measurements were made in the turbulence generated by a square-mesh grid at four different Reynolds numbers, and on the central plane of the turbulent wake behind a cylinder at two different Reynolds numbers. The variation of the results over all six cases is small enough to encourage the belief (based on the universal equilibrium theory) that the limiting values for very large Reynolds number are independent of the large-scale properties of the turbulence. The interesting aspect of the results is the rapid increase of the flatness factor (8.1.3) with n. The probability distributions of $\partial^2 u_1/\partial x_1^2$ and $\partial^3 u_1/\partial x_1^3$ must be very different from a normal distribution to have flatness factors in the neighbourhood of 5 and 6 respectively. The interpretation of these remarkable observations will be considered in §8.4.

8.2. The hypothesis of a normal distribution of the velocity field associated with the energy-containing eddies

Since the velocity at any point is subject to the influence of a large number of random eddies or flow patterns in its neighbourhood, we might expect, from a rough use of the Central Limit Theorem, that this velocity has an approximately normal probability distribution. The energy-containing eddies have their origin in some mechanical

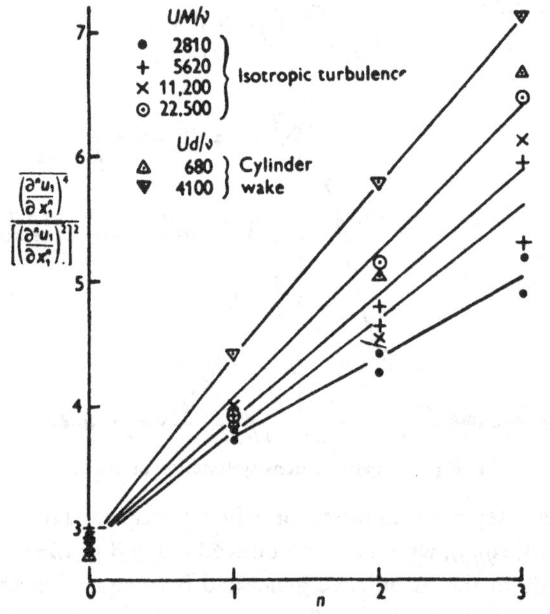

Fig. 8.5. Flatness factor of velocity derivatives (from Batchelor and Townsend, 1949).

stirring action, such as the passage of air through an array of rods, and it would indeed be a little surprising if these eddies did not give rise to a velocity (at one point) distributed approximately according to an error law. The same argument would not serve to show that the joint distribution of the velocities at two points is normal, since the relation between the velocities at the two points must conform to the equations of motion and of continuity, and these are unlikely to permit a distribution of the pure chance type. The inertia terms of the equation of motion in particular will exercise a strong influence on the relation between the two velocities and will lead to such effects as a difference between the probability distribution of positive relative velocities (extension of the fluid between the two points) and the distribution of negative relative velocities (contraction between the two points).

When the two points are close together, the relation between the velocities at the two points is very strong, and the difference between the velocities will have a probability distribution which is governed to a large extent by the Navier-Stokes equation. As we saw earlier, the difference between the velocities at two neighbouring points is determined by the Fourier coefficients corresponding to the small eddies. Eddies smaller than a certain size owe their existence entirely to the non-linear transfer down the spectrum, and the smaller the eddy the more prolonged, so to speak, has been the influence of the non-linear terms. We may expect the statistical characteristics of the small eddies to reflect this influence, and the distribution of the velocity difference $(\mathbf{u}'-\mathbf{u})$ will probably depart furthest from a normal distribution at very small values of r.

These general expectations are consistent with the measurements described in the previous section. The distribution of the velocity at one point was found to be fairly accurately normal. The distribution of $u_1'-u_1$ was found to be accurately normal if r is large enough, but is significantly different from a normal distribution at small values of r/l, the maximum difference occurring at $r=0$. As r is decreased, the skewness factor of the distribution of $u_1'-u_1$ becomes different from zero considerably faster than the flatness factor becomes different from 3·0. This is also understandable in view of the interpretation of the triple correlation (of which $\overline{(\partial u_1/\partial x_1)^3}$ is a special form) as a measure of the inertial transfer of energy, which is significant for both large and small eddies.

Assuming that the behaviour of the above third- and fourth-order product mean values is typical of the behaviour of higher order products, we are led to the working hypothesis that the part of the probability distribution of the function $\mathbf{u}(\mathbf{x})$ that is determined by the motion associated with the energy-containing eddies is approximately normal, at any rate so far as the values of the velocity at no more than two points are concerned. This is an approximation which will be better for some purposes than for others; the measurements suggest it will give reasonably accurate predictions about the relation between the fourth- and second-order two-point mean values but less accurate predictions about the triple-velocity correlation. The potential usefulness of the hypothesis is obviously very great, since it provides a method of expressing complicated velocity-product mean values in terms of the fundamental correlation $R_{ij}(\mathbf{r})$, but it must be used with caution in view of the limited scope of the supporting experimental data. There is no theoretical basis for the supposition that departures from a normal distribution first become appreciable at the *end* of the energy-containing range of eddy sizes, so that this part of the hypothesis must stand or fall on its empirical merits.

Another statement of the hypothesis may be made in terms of the Fourier coefficients $d\mathbf{Z}(\boldsymbol{\kappa})$. It was stated in § 2.5 that the velocity could be represented in the form

$$\mathbf{u}(\mathbf{x}) = \int e^{i\boldsymbol{\kappa}.\mathbf{x}}\, d\mathbf{Z}(\boldsymbol{\kappa}), \tag{8.2.1}$$

where the values of the random increments $d\mathbf{Z}(\boldsymbol{\kappa})$ for different values of $\boldsymbol{\kappa}$ are uncorrelated. $d\mathbf{Z}(\boldsymbol{\kappa})$ is a linear functional of the velocity $\mathbf{u}(\mathbf{x})$ (see (2.5.2)), and it follows† that if the function $\mathbf{u}(\mathbf{x})$ is normally distributed, so too is the function $d\mathbf{Z}(\boldsymbol{\kappa})$. But if the values of $d\mathbf{Z}(\boldsymbol{\kappa})$ at any two values of $\boldsymbol{\kappa}$ are normally distributed and uncorrelated, they must also be statistically independent. Hence a normal distribution of the function $\mathbf{u}(\mathbf{x})$ requires the values of $d\mathbf{Z}(\boldsymbol{\kappa})$ for different values of $\boldsymbol{\kappa}$ to be independent.

It is also of interest to inquire if the converse can be established by applying the Central Limit Theorem to the (effective) sum on the right side of (8.2.1). Normality of the distribution of $\mathbf{u}(\mathbf{x})$ does not follow immediately from independence of the Fourier coefficients; for instance, if the set $d\mathbf{Z}(\boldsymbol{\kappa})$ consisted of a set of independent jumps at the random points $\boldsymbol{\kappa}_1, \boldsymbol{\kappa}_2, \ldots$ which varied from one realization to another (these conditions being theoretically compatible with a continuous spectrum), the Central Limit Theorem would not be applicable. However, we can invoke further conditions which rule

† H. Cramér, *Mathematical methods of statistics*, Princeton University Press, 1946, chap. 24.

out such rather special cases. Provided the velocity field is such that the ergodic property holds, the requirement of a continuous spectrum means that the spectrum obtained from (almost) every realization (by a space average) must be continuous, which excludes the above case. Roughly speaking, the ergodic property demands that for (almost) every realization of the velocity field there must be an indefinitely large number of Fourier components in any interval $\delta\boldsymbol{\kappa}$, and this will now ensure, by the Central Limit Theorem, that the sum of a large number of terms like $\int_{\delta\kappa} e^{i\boldsymbol{\kappa}\cdot\mathbf{x}}\,d\mathbf{Z}(\boldsymbol{\kappa})$ is normally distributed when the values of $d\mathbf{Z}(\boldsymbol{\kappa})$ for different $\boldsymbol{\kappa}$ are independent.†

The hypothesis of a normal distribution of the energy-containing eddies is thus equivalent to the hypothesis of independence of the Fourier coefficients $d\mathbf{Z}(\boldsymbol{\kappa})$ over the appropriate range of wave-numbers. Put in the latter form, the hypothesis can be made more precise; we can suppose that the $d\mathbf{Z}(\boldsymbol{\kappa})$ are independent for $\kappa < \sigma$, where σ is to be determined empirically and is expected (from the existing data) to be such that $\int_0^\sigma E(\kappa)\,d\kappa$ is an appreciable fraction of the total energy.

Use of the above hypothesis was first made by M. Millionshtchikov (1941 *a*, *b*), who assumed that fourth-order and second-order two-point product mean values were related in the same way as for a normal distribution of $\mathbf{u}(\mathbf{x})$, for a particular purpose which will not be described here. W. Heisenberg (1948 *a*) assumed independence of the coefficients $d\mathbf{Z}(\boldsymbol{\kappa})$ in order to get an estimate of the covariance of the pressure fluctuations at two points, and a similar use of the hypothesis was made later by A. M. Obukhoff (1949*a*) and G. K. Batchelor (1951). (This particular application of the hypothesis will be described in the next section.) A related use of the hypothesis is being made in some of the current work on density fluctuations in a fluid in turbulent motion.

8.3. Determination of the pressure covariance

In quite a number of physical problems in which turbulent motion occurs, the fluctuating pressure is the quantity of most importance. An obvious example is the determination of the sound

† I am grateful to Professor M. S. Bartlett for his help with the preparation of these two paragraphs.

field generated by the turbulent motion when the fluid is compressible. We shall not consider these problems here, but will show how some of the statistical characteristics of the pressure field in isotropic turbulence can be determined with the aid of the hypothesis of the preceding section.

The divergence of the Navier-Stokes equation yields the following equation for the pressure distribution:

$$\frac{1}{\rho}\nabla^2 p(\mathbf{x}) = -\frac{\partial^2 u_i u_j}{\partial x_i \partial x_j}. \tag{8.3.1}$$

From this and a similar equation at $\mathbf{x}'$ (where the pressure and velocity are p' and $\mathbf{u}'$) we find

$$\frac{1}{\rho^2}\nabla^2_{\mathbf{x}} p \,.\, \nabla^2_{\mathbf{x}'} p' = \frac{\partial^2 u_i u_j}{\partial x_i \partial x_j}\frac{\partial^2 u'_l u'_m}{\partial x'_l \partial x'_m} = \frac{\partial^4 u_i u_j u'_l u'_m}{\partial x_i \partial x_j \partial x'_l \partial x'_m},$$

which gives the equation for the pressure covariance as

$$\frac{1}{\rho^2}\nabla^4 \overline{pp'} = \frac{\partial^4 \overline{u_i u_j u'_l u'_m}}{\partial r_i \partial r_j \partial r_l \partial r_m}, \tag{8.3.2}$$

where $\mathbf{r} = \mathbf{x}' - \mathbf{x}$, and ∇^2 now has the meaning $\partial^2/\partial r_i \partial r_i$. When the turbulence is isotropic the quantities on both sides of (8.3.2) are functions of r^2 alone, and we can write

$$\frac{1}{\rho^2}\overline{pp'} = P(r), \qquad \frac{\partial^4 \overline{u_i u_j u'_l u'_m}}{\partial r_i \partial r_j \partial r_l \partial r_m} = W(r), \tag{8.3.3}$$

in which case (8.3.2) becomes

$$W(r) = \left(\frac{\mathrm{d}^2}{\mathrm{d}r^2} + \frac{2}{r}\frac{\mathrm{d}}{\mathrm{d}r}\right)^2 P(r)$$

$$= \left(\frac{\mathrm{d}^4}{\mathrm{d}r^4} + \frac{4}{r}\frac{\mathrm{d}^3}{\mathrm{d}r^3}\right) P(r) = \frac{1}{r}\frac{\mathrm{d}^4 rP(r)}{\mathrm{d}r^4}. \tag{8.3.4}$$

Provided we can assume that $P(r)$ and $W(r)$ approach zero with sufficient rapidity as $r \to \infty$, the explicit expression for the pressure covariance in terms of the fourth-order velocity-product mean value is

$$P(r) = \frac{1}{6r}\int_r^\infty y(y-r)^3\, W(y)\,\mathrm{d}y. \tag{8.3.5}$$

A corresponding equation for the pressure 'spectrum' function defined by

$$\Pi(\kappa) = \frac{1}{8\pi^3}\int P(r)\,e^{-i\boldsymbol{\kappa}.\mathbf{r}}\,d\mathbf{r} = \frac{1}{2\pi^2\kappa^2}\int_0^\infty P(r)\,\kappa r \sin \kappa r\, dr \tag{8.3.6}$$

can also be obtained. It is not possible to define a Fourier transform of $\overline{u_i u_j u'_l u'_m}$, since it tends to the non-zero value $\overline{u_i u_j}.\overline{u'_l u'_m}$ as $r\to\infty$, but we can define a tensor $\Omega_{ijlm}(\boldsymbol{\kappa})$ by the relation

$$\overline{u_i u_j u'_l u'_m} - \overline{u_i u_j}.\overline{u'_l u'_m} = \int \Omega_{ijlm}(\boldsymbol{\kappa})\,e^{i\boldsymbol{\kappa}.\mathbf{r}}\,d\boldsymbol{\kappa}, \tag{8.3.7}$$

whence we have, from (8.3.3),

$$W(r) = \int \kappa_i\kappa_j\kappa_l\kappa_m \Omega_{ijlm}(\boldsymbol{\kappa})\,e^{i\boldsymbol{\kappa}.\mathbf{r}}\,d\boldsymbol{\kappa}. \tag{8.3.8}$$

We then find, from (8.3.4) and the relation inverse to (8.3.6), that the pressure spectrum is given by

$$\Pi(\kappa) = \kappa^{-4}\kappa_i\kappa_j\kappa_l\kappa_m \Omega_{ijlm}(\boldsymbol{\kappa}). \tag{8.3.9}$$

This is the stage at which it is necessary to introduce some approximation or hypothesis, since we have little theoretical or experimental information about the fourth-order two-point velocity-product mean value. The hypothesis described in the preceding section fills the need very well, since it supplies a simple relation between fourth-order product mean values and second-order product mean values, the latter of which can be regarded as 'known' in view of the intensive investigation to which they have been subjected. When the joint-probability distribution of $\mathbf{u}$ and $\mathbf{u}'$ is normal, the probability density function $Q(\mathbf{u}, \mathbf{u}')$ and the moment-generating function $M(\boldsymbol{\alpha}, \boldsymbol{\beta})$ are given by

$$M(\boldsymbol{\alpha}, \boldsymbol{\beta}) = \iint Q(\mathbf{u}, \mathbf{u}')\,e^{(\boldsymbol{\alpha}.\mathbf{u}+\boldsymbol{\beta}.\mathbf{u}')}\,d\mathbf{u}\,d\mathbf{u}'$$

$$= \exp\left[\tfrac{1}{2}(\overline{u_l u_m}.\alpha_l\alpha_m + 2\overline{u_l u'_m}.\alpha_l\beta_m + \overline{u'_l u'_m}.\beta_l\beta_m)\right], \tag{8.3.10}$$

from which it follows that

$$\overline{u_i u_j u'_l u'_m} = \overline{u_i u_j}.\overline{u'_l u'_m} + \overline{u_i u'_l}.\overline{u_j u'_m} + \overline{u_i u'_m}.\overline{u_j u'_l}. \tag{8.3.11}$$

The use of this relation will now permit us to determine $P(r)$ in terms of $R_{ij}(\mathbf{r})$ and thence in terms of the basic scalar function

$u^2 f(r)$, but we must keep in mind that the approximation (8.3.10) is known to be accurate only for quantities determined by the energy-containing eddies.

Substituting (8.3.11) in (8.3.3) and making use of the equation of continuity in the form (2.4.6), we find

$$W(r) = 2\frac{\partial^2 R_{il}(\mathbf{r})}{\partial r_j \partial r_m}\frac{\partial^2 R_{jm}(\mathbf{r})}{\partial r_i \partial r_l}. \qquad (8.3.12)$$

When the turbulence is isotropic (as it is in the case in which measurements have led to the hypothesis of a normal distribution) $R_{ij}(\mathbf{r})$ can be expressed in terms of the single scalar function $u^2 f(r)$ by means of the relations (3.4.5) and (3.4.6), whence after some elementary algebra the expression for $W(r)$ becomes

$$W(r) = 4u^4\left(2f''^2 + 2f'f''' + \frac{10}{r}f'f'' + \frac{3}{r^2}f'^2\right), \qquad (8.3.13)$$

where dashes denote differentiation with respect to r. The expression (8.3.5) for the pressure covariance then reduces to

$$P(r) = 2u^4\int_r^\infty \left(y - \frac{r^2}{y}\right)[f'(y)]^2\,dy. \qquad (8.3.14)$$

In particular, the mean-square pressure fluctuation is

$$\frac{1}{\rho^2}\overline{p^2} = P(0) = 2u^4\int_0^\infty yf'^2\,dy, \qquad (8.3.15)$$

and the mean square pressure gradient is

$$\frac{1}{\rho^2}\overline{(\nabla p)^2} = -3P_0'' = 12u^4\int_0^\infty \frac{1}{y}f'^2\,dy. \qquad (8.3.16)$$

Corresponding to (8.3.14) there is a relation between the spectrum functions $\Pi(\kappa)$ and $E(\kappa)$ which is found (see Batchelor, 1951) to be

$$\Pi(\kappa) = \frac{1}{8\pi^2}\int E(\kappa')\,E(|\boldsymbol{\kappa} - \boldsymbol{\kappa}'|)\frac{\sin^4\theta}{|\boldsymbol{\kappa} - \boldsymbol{\kappa}'|^4}\,d\boldsymbol{\kappa}', \qquad (8.3.17)$$

where θ is the angle between $\boldsymbol{\kappa}$ and $\boldsymbol{\kappa}'$. It is clear from the form of the integrand that the value of $\Pi(\kappa)$ at $\kappa = \kappa_0$, say, is determined by the form of the function $E(\kappa)$ for values of κ which are of the same order of magnitude as κ_0. Consequently, if we can infer from the experimental results described in §8.1 that the hypothesis of independence of the increments $d\mathbf{Z}(\boldsymbol{\kappa})$ is accurate for $\kappa < \sigma$, (8.3.17) is a valid approximate formula for $\Pi(\kappa)$ provided $\kappa < \sigma$ (although the accuracy may not be very high for values of κ *near* σ in view of the contributions to the integral (8.3.17) from eddies not included in the hypothesis). The mean-square pressure fluctuation follows from (8.3.17) as

$$\begin{aligned}\frac{\overline{p^2}}{\rho^2} &= \int \Pi(\kappa)\, d\boldsymbol{\kappa} \\ &= \frac{1}{8\pi^2}\iint E(\kappa')\,E(\kappa'')\frac{\sin^4\phi}{|\boldsymbol{\kappa}'-\boldsymbol{\kappa}''|^4}\, d\boldsymbol{\kappa}'\, d\boldsymbol{\kappa}'', \end{aligned} \qquad (8.3.18)$$

where $\boldsymbol{\kappa}'' = \boldsymbol{\kappa}' - \boldsymbol{\kappa}$, and ϕ is the angle between $\boldsymbol{\kappa}'$ and $\boldsymbol{\kappa}''$. The integrations over the directions of the vectors $\boldsymbol{\kappa}'$ and $\boldsymbol{\kappa}''$ can be carried out, giving

$$\frac{\overline{p^2}}{\rho^2} = \int_0^\infty \int_0^\infty E(\kappa')\,E(\kappa'')\,I\left(\frac{\kappa'}{\kappa''}\right) d\kappa'\, d\kappa'', \qquad (8.3.19)$$

where

$$I(s) = I\left(\frac{1}{s}\right) = \tfrac{1}{2}(s^2+s^{-2}) - \tfrac{1}{3} - \tfrac{1}{4}(s+s^{-1})(s-s^{-1})^2 \log\frac{1+s}{|1-s|}.$$

The weighting factor $I(s)$ is such that when s is of order unity, $I(s)$ is of order unity, and when $s \gg 1$, $I(s) \approx \frac{16}{15}s^{-2}$. Consequently $\overline{p^2}/\rho^2$ is determined, according to (8.3.19), principally by the range of wave-numbers near the position of the maximum of $E(\kappa)$; if the Fourier coefficients determining the energy-containing range of the spectrum are statistically independent, (8.3.19) is a valid approximate formula and $\overline{p^2}/\rho^2$ will be determined by the form of $E(\kappa)$ in that range. It is a corollary that the order of magnitude of $\overline{p^2}/\rho^2$ is that of $\left[\int_0^\infty E(\kappa)\, d\kappa\right]^2$, that is, of $\frac{1}{4}(\overline{\mathbf{u}^2})^2$.

Likewise the expression for the mean-square pressure gradient in terms of the energy spectrum function (Heisenberg, 1948*a*) is

$$\begin{aligned}\frac{1}{\rho^2}\overline{(\nabla p)^2} &= \int \kappa^2 \Pi(\kappa)\, d\boldsymbol{\kappa} \\ &= \frac{1}{8\pi^2}\iint E(\kappa')\,E(\kappa'')\frac{\sin^4\phi}{|\boldsymbol{\kappa}'-\boldsymbol{\kappa}''|^2}\, d\boldsymbol{\kappa}'\, d\boldsymbol{\kappa}'' \\ &= \int_0^\infty\int_0^\infty E(\kappa')\,E(\kappa'')\,\kappa'\kappa'' J\left(\frac{\kappa'}{\kappa''}\right) d\kappa'\, d\kappa'', \end{aligned} \qquad (8.3.20)$$

where

$$J(s)=J\left(\frac{1}{s}\right)=-\tfrac{1}{8}(s^3+s^{-3})+\tfrac{11}{24}(s+s^{-1})+\tfrac{1}{16}(s-s^{-1})^4\log\frac{1+s}{|1-s|}.$$

The above remarks about the behaviour of $I(s)$ apply also to $J(s)$, so that if the Fourier coefficients for wave-numbers near the position of the maximum of $\kappa E(\kappa)$ are statistically independent of each other, (8.3.20) is a valid approximate formula and $\frac{1}{\rho^2}\overline{(\nabla p)^2}$ will be determined by the form of $E(\kappa)$ in that range. This is a condition which has not yet been shown experimentally to be satisfied, and the evidence described in §8.1 suggests that at best it will be a fair approximation only. We should therefore treat the deductions about $\frac{1}{\rho^2}\overline{(\nabla p)^2}$ $\left(\text{and about } \frac{1}{\rho^2}(\overline{p^2}-\overline{pp'}) \text{ for small values of } r\right)$ with more reserve than those about $\frac{1}{\rho^2}\overline{p^2}$.

If information about the function $u^2f(r)$ or $E(\kappa)$ is available, we can use the above formulae to calculate the pressure covariance or the spectrum function $\Pi(\kappa)$. It has been remarked (in §7.2) that when the Reynolds number of the turbulence is only moderately large (moderate from an experimental point of view) the functions $u^2f(r)$ and $E(\kappa)$ have reached their asymptotic shapes (i.e. the shapes for infinite Reynolds number) over the ranges of r and κ associated with the energy-containing eddies. This asymptotic shape is therefore an important member of the family of velocity correlation functions, and the appropriate calculations of $P(r)$ from the relation (8.3.14) have been made (Batchelor, 1951) with the result shown in fig. 8.6. The calculation also showed that

$$\overline{p^2}=0{\cdot}34\rho^2u^4=0{\cdot}15(\tfrac{1}{2}\rho\overline{\mathbf{u}^2})^2. \tag{8.3.21}$$

The mean-square pressure gradient, on the other hand, depends significantly on the form of $E(\kappa)$ outside the energy-containing range, which is not known accurately at these moderate Reynolds numbers. However, provided the Reynolds number is so large that an inertial subrange of the spectrum exists, the value of $\frac{1}{\rho^2}\overline{(\nabla p)^2}$ depends chiefly on the extent of this range, and one finds the following approximate formula:

$$\frac{1}{\rho^2}\overline{(\nabla p)^2} = 226\frac{\nu u^3}{\lambda^3}. \tag{8.3.22}$$

In this latter case of very large Reynolds number, the value of $\frac{1}{\rho^2}(\overline{p^2}-\overline{pp'})$ for small values of r is determined by the inertial sub-range and the appropriate form of (8.3.14) is the very simple formula

$$\frac{1}{\rho^2}\overline{(p'-p)^2} = \{\overline{[u_1(x_1+r,x_2,x_3)-u_1(x_1,x_2,x_3)]^2}\}^2, \tag{8.3.23}$$

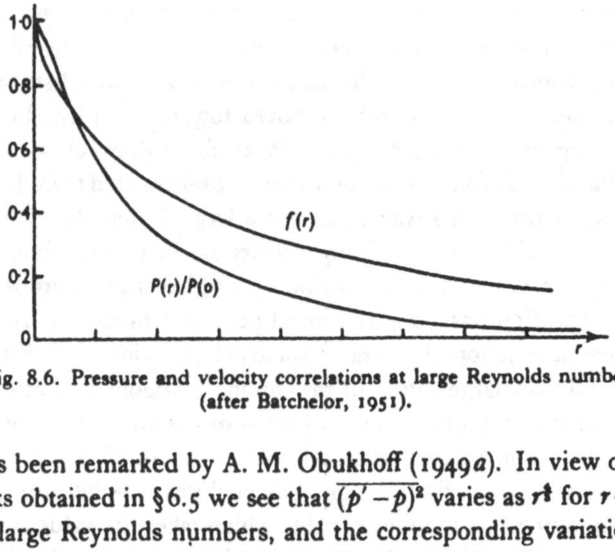

Fig. 8.6. Pressure and velocity correlations at large Reynolds number (after Batchelor, 1951).

as has been remarked by A. M. Obukhoff (1949*a*). In view of the results obtained in §6.5 we see that $\overline{(p'-p)^2}$ varies as $r^{4/3}$ for $r \ll l$ at very large Reynolds numbers, and the corresponding variation of the spectrum function $\Pi(\kappa)$ is readily found to be $\Pi(\kappa) \propto \kappa^{-7/3}$ for $\kappa \gg 1/l$.

8.4. The small-scale properties of the motion

Our understanding of the general character of the small-scale features of turbulent motion is very far from complete, but the problems concerned are so interesting that a few tentative remarks will be made here. Very few theoretical or experimental results have been established so that for the most part we must proceed by analogy and plausible inference. A very striking observation concerning the small-scale properties of the motion has already been described in § 8.1, and we can take this as a starting point for the discussion. The measurements described in fig. 8.5 show that the flatness factor of the probability distribution of the various velocity derivatives increases quite steeply with the order of the derivative, and the flatness factors of the distributions of u_1, $\partial u_1/\partial x_1$, $\partial^2 u_1/\partial x_1^2$ and $\partial^3 u_1/\partial x_1^3$ were found to be in the neighbourhood of 3, 4, 5 and 6 respectively. Since the representation of $\partial^n u_1/\partial x_1^n$ in terms of Fourier components is

$$\frac{\partial^n u_1}{\partial x_1^n} = \iota^n \int e^{\iota \boldsymbol{\kappa}.\mathbf{x}} \kappa_1^n \, dZ_1(\boldsymbol{\kappa}),$$

the measurements show, in effect, that the contribution to $\mathbf{u}(\mathbf{x})$ from a given range of wave-numbers has a probability distribution which becomes more markedly different from a normal distribution as that range of wave-numbers moves towards infinity. Equivalently, the Fourier coefficients $d\mathbf{Z}(\boldsymbol{\kappa})$ for values of κ in the neighbourhood of κ_0, say, become more closely related to each other statistically (in such a way as to give a large flatness factor to the distribution of the corresponding velocity derivative) as κ_0 increases.

A large flatness factor of a distribution implies that the probability density function has a higher central peak and broader skirts than a Gaussian function of the same standard deviation, and that very small and very large values of the random variable are both more probable than for a normal distribution of the same deviation. An extreme version of such a distribution is obtained if the central peak becomes of infinite height but encloses a finite area $1-\gamma$, corresponding to a random variable which takes the value zero for a fraction $1-\gamma$ of the total number of realizations and is distributed in some given way for the remainder of the realizations. If we make the simple assumption that $\partial^n u_1/\partial x_1^n$ is distributed in this extreme way, with a normal distribution of the non-zero values, we have

$$\overline{\left(\frac{\partial^n u_1}{\partial x_1^n}\right)^4} \Big/ \left[\overline{\left(\frac{\partial^n u_1}{\partial x_1^n}\right)^2}\right]^2 = \frac{3\cdot 0}{\gamma}, \qquad (8.4.1)$$

so that the measured flatness factors for $n = 0, 1, 2, 3$ (average values of which are 3·0, 3·9, 4·9, 5·9 respectively), correspond to $\gamma = 1\cdot 0$, 0·77, 0·61, 0·51 respectively. These values of γ are effective 'intermittency' factors in the sense that they describe the fraction of the total number of realizations for which the derivatives (at a given point and time) fluctuate and take non-zero values; or, assuming the equivalence of probability and space averages, they describe the fraction of the total space of one realized velocity field for which the derivatives are non-zero, if the distributions have the above hypothetical form.

Direct observation of the variation of u_1, $\partial u_1/\partial x_1$, $\partial^2 u_1/\partial x_1^2$ and $\partial^3 u_1/\partial x_1^3$ with x_1, as shown on the screen of an oscillograph, suggests that the above possible interpretation of the high measured flatness factors is at least a qualitative approximation to the truth (Batchelor and Townsend, 1949). These visual observations of the traces of $\partial^2 u_1/\partial x_1^2$ and $\partial^3 u_1/\partial x_1^3$ revealed a fairly definite alternation between periods of quiescence, during which the magnitude of the derivative is small, and periods of activity during which the derivative fluctuates in an apparently random fashion. (Visual observations of the trace do not, of course, provide information about whether the derivative is normally distributed during periods of activity.) The periods or regions of activity were large enough to contain a considerable number of oscillations of the derivative. These properties of the higher derivatives were observed both for turbulence generated by a grid and for the turbulence in the wake of a long cylinder, and are evidently intrinsic to the micro-structure of turbulence in general—as, indeed, we should expect at high Reynolds numbers.

The inference, then, is that there is an uneven distribution, in space, of the energy associated with the large wave-number components of the turbulence, and that the higher the wave-number, the more does the associated energy tend to occur in confined regions of space (meaning that if a Fourier resolution of the velocity field *within* a region of activity were made, the amplitude of the component at the relevant wave-number would be found to be large, while a Fourier resolution of the velocity field within a region of quiescence would give a very small amplitude; the amplitude of the component for the field as a whole will lie somewhere between these two amplitudes). There are other hydrodynamical situations in which a similar tendency appears. It is well known that certain steady motions are unstable to small disturbances and tend to a state in which the vorticity is concentrated in isolated regions. Direct observations and calculations† both show that a plane vortex sheet, i.e. a layer across which the velocity is discontinuous, is unstable to small two-dimensional disturbances and tends to break up into a row of closely wound spirals. These spirals are activated regions for the vorticity in the sense of the interpretation of the observations of turbulence. It is a plausible speculation that flows such that higher order derivatives ($\nabla \times (\nabla \times \mathbf{u})$, etc.) are uniform over a plane and zero elsewhere are similarly unstable, and give rise to a number of regions of concentration of the derivatives.

A related property of hydrodynamical systems at large Reynolds number is the tendency for regions of rapid change of the velocity to form. This important tendency was mentioned in Chapter VI as an illustration of the way in which energy is transferred from small to large wave-number harmonic components. What is relevant to the present context is the tendency for a small number of strong discontinuities—rather than small discontinuities distributed uniformly throughout the fluid—to form, giving another expression of the tendency for the energy associated with Fourier components of large wave-number to have an uneven spatial distribution. J. M. Burgers (1948*b*, 1950*a*, *b*) has found that for a hypothetical model system which satisfies a simplified equation of motion, the appearance of a few strong discontinuities (or near-discontinuities, owing to the smoothing effect of viscosity) can be explained as the result of a gradual overtaking and coalescence of the many small discontinuities which are first formed, and it seems very probable that a similar process occurs in the real system.

On the whole, the apparent spottiness of the spatial distribution of the energy of high wave-number components is qualitatively in agreement with our general ideas (which admittedly are still very crude) about the effect of the non-linear inertia forces. A case which offers us a better chance of examining the above effects is that of turbulent motion in two dimensions. Motion in two dimensions has the simple property that the vorticity of a fluid element is unchanged, except by molecular diffusion, as the element follows the motion (and precisely because of this special characteristic of two-

† L. Rosenhead, 'The formation of vortices from a surface of discontinuity', *Proc. Roy. Soc.* A, **134**, 1932, 170.

dimensional motion we should beware of assuming too close a relation between two- and three-dimensional turbulence). Hence for a motion with zero viscosity, the integrals

$$\int_0^\infty E(\kappa)\,d\kappa, \quad \int_0^\infty \kappa^2 E(\kappa)\,d\kappa$$

are constant; even when ν is finite but small the integrals will be constant until such time as energy has been transferred to high wave-numbers at which viscous forces are significant. The effect of the non-linear term of the equation will be to transfer energy over an increasingly wide range of wave-numbers, and if we imagine the initial state to be such that all the energy lies in the range $0 < \kappa < \kappa'$, one of the effects of the non-linear term will be to transfer energy to wave-numbers $\kappa > \kappa'$. But if there is a transfer of energy across $\kappa = \kappa'$, the constancy of $\int_0^\infty \kappa^2 E(\kappa)\,d\kappa$ demands that there should be an even greater flow of energy in the opposite direction within the range $0 < \kappa < \kappa'$. The first moment $\int_0^\infty \kappa E(\kappa)\,d\kappa$ thus becomes smaller and smaller, while the wave-number which divides the region in which energy is flowing to high wave-numbers from that in which there is a flow in the reverse direction, continually increases. This net tendency for the bulk of the energy to concentrate in the small wave-numbers means that fluid elements with similarly signed vorticity must tend to group together; in no other way is it possible for the scale of the velocity distribution to increase. We expect, therefore, that from the original motion there will gradually emerge a few strong isolated vortices and that vortices of the same sign will continue to tend to group together. The differences between this motion and three-dimensional turbulence are very great, but the above argument suggests they have in common the property that the fluctuations in the velocity derivatives tend to occur in confined regions of space.

L. Onsager (1949) has arrived at a similar conclusion about the tendency for a small number of strong isolated vortices to form in a two-dimensional motion consisting of a random distribution of line vortices, from an argument based on the methods of statistical mechanics.

BIBLIOGRAPHY OF RESEARCH ON HOMOGENEOUS TURBULENCE

AGOSTINI, L. (1949*a*). La fonction spectrale de la turbulence isotrope. *C.R. Acad. Sci., Paris*, **228**, 736.

AGOSTINI, L. (1949*b*). Sur quelques propriétés de la fonction de corrélation totale. *C.R. Acad. Sci., Paris*, **228**, 810.

AGOSTINI, L. and BASS, J. (1950). Les théories de la turbulence. *Publ. Sci. Tech. Ministère l'Air*, no. 237.

BASS, J. (1946). Les méthodes modernes du calcul des probabilités et leur application au problème de la turbulence. *G.R.A. Rapport Tech.* no. 228.

BASS, J. (1948). Les bases d'une théorie statistique de la turbulence. *Proc. 7th Int. Congr. Appl. Mech.* **2**, 212.

BASS, J. (1949). Sur les bases mathématiques de la théorie de la turbulence d'Heisenberg. *C.R. Acad. Sci., Paris*, **228**, 228.

BATCHELOR, G. K. (1946). The theory of axisymmetric turbulence. *Proc. Roy. Soc.* A, **186**, 480.

BATCHELOR, G. K. (1947). Kolmogoroff's theory of locally isotropic turbulence. *Proc. Camb. Phil. Soc.* **43**, 533.

BATCHELOR, G. K. (1948*a*). Energy decay and self-preserving correlation functions in isotropic turbulence. *Quart. Appl. Math.* **6**, 97.

BATCHELOR, G. K. (1948*b*). Recent developments in turbulence research. *Proc. 7th Int. Congr. Appl. Mech.* introd. vol.

BATCHELOR, G. K. (1949*a*). The role of big eddies in homogeneous turbulence. *Proc. Roy. Soc.* A, **195**, 513.

BATCHELOR, G. K. (1949*b*). Diffusion in a field of homogeneous turbulence. I. Eulerian analysis. *Aust. J. Sci. Res.* **2**, 437.

BATCHELOR, G. K. (1950*a*). The application of the similarity theory of turbulence to atmospheric diffusion. *Quart. J. R. Met. Soc.* **76**, 133.

BATCHELOR, G. K. (1950*b*). On the spontaneous magnetic field in a conducting liquid in turbulent motion. *Proc. Roy. Soc.* A, **201**, 405.

BATCHELOR, G. K. (1951). Pressure fluctuations in isotropic turbulence. *Proc. Camb. Phil. Soc.* **47**, 359.

BATCHELOR, G. K. (1952*a*). Diffusion in a field of homogeneous turbulence. II. The relative motion of particles. *Proc. Camb. Phil. Soc.* **48**, 345.

BATCHELOR, G. K. (1952*b*). Turbulent motion. *Rep. Progr. Phys.* **15**, 101.

BATCHELOR, G. K. (1952*c*). The effect of homogeneous turbulence on material lines and surfaces. *Proc. Roy. Soc.* A, **213**, 349.

BATCHELOR, G. K. and STEWART, R. W. (1950). Anisotropy of the spectrum of turbulence at small wave-numbers. *Quart. J. Mech. Appl. Math.* **3**, 1.

BATCHELOR, G. K. and TOWNSEND, A. A. (1947). Decay of vorticity in isotropic turbulence. *Proc. Roy. Soc.* A, **190**, 534.

BATCHELOR, G. K. and TOWNSEND, A. A. (1948*a*). Decay of isotropic turbulence in the initial period. *Proc. Roy. Soc.* A, **193**, 539.

BATCHELOR, G. K. and TOWNSEND, A. A. (1948*b*). Decay of turbulence in the final period. *Proc. Roy. Soc.* A, **194**, 527.

BATCHELOR, G. K. and TOWNSEND, A. A. (1949). The nature of turbulent motion at large wave-numbers. *Proc. Roy. Soc.* A, **199**, 238.

BETCHOV, R. (1948). L'analyse spectrale de la turbulence. *Proc. Acad. Sci. Amst.* **51**, 1063.

BURGERS, J. M. (1939). Mathematical examples illustrating relations occurring in the theory of turbulent fluid motion. *Verh. K. Akad. Wet. Amst.*, Addeel. Nat. (1st Section), **17**, 1.

BURGERS, J. M. (1940*a*). Application of a model system to illustrate some points of the statistical theory of free turbulence. *Proc. Acad. Sci. Amst.* **43**, 2.

BURGERS, J. M. (1940*b*). On the application of statistical mechanics to the theory of turbulent fluid motion. *Proc. Acad. Sci. Amst.* **43**, 936, 1153.

BURGERS, J. M. (1941*a*). Beschouwingen over de statistische theorie der turbulente stroming. *Nederl. Tijdschr. Natuurkunde*, **8**, 5.

BURGERS, J. M. (1941*b*). On the distinction between irregular and systematic motion in diffusion problems. *Proc. Acad. Sci. Amst.* **44**, 344.

BURGERS, J. M. (1948*a*). A mathematical model illustrating the theory of turbulence. *Adv. Appl. Mech.* **1**, 171.

BURGERS, J. M. (1948*b*). Spectral analysis of an irregular function. *Proc. Acad. Sci. Amst.* **51**, 1073.

BURGERS, J. M. (1950*a*). The formation of vortex sheets in a simplified type of turbulent motion. *Proc. Acad. Sci. Amst.* **53**, 122.

BURGERS, J. M. (1950*b*). Correlation problems in a one-dimensional model of turbulence. I–IV. *Proc. Acad. Sci. Amst.* **53**, 247, 393, 718, 732.

CHANDRASEKHAR, S. (1949*a*). On Heisenberg's elementary theory of turbulence. *Proc. Roy. Soc.* A, **200**, 20.

CHANDRASEKHAR, S. (1949*b*). Turbulence—a physical theory of astrophysical interest. *Astrophys. J.* **110**, 329.

CHANDRASEKHAR, S. (1950*a*). The theory of axisymmetric turbulence. *Philos. Trans.* A, **242**, 557.

CHANDRASEKHAR, S. (1950*b*). The decay of axisymmetric turbulence. *Proc. Roy. Soc.* A, **203**, 358.

CHANDRASEKHAR, S. (1951*a*). The invariant theory of isotropic turbulence in magneto-hydrodynamics. *Proc. Roy. Soc.* A, **204**, 435.

CHANDRASEKHAR, S. (1951*b*). The invariant theory of isotropic turbulence in magneto-hydrodynamics. Part II. *Proc. Roy. Soc.* A. **207**, 306.

CHANDRASEKHAR, S. (1953). Some aspects of the statistical theory of turbulence. *Proc. 4th Symp. in Appl. Math.*, Amer. Math. Soc., p. 1.

CHOU, P. Y. (1945). On velocity correlations and the solutions of the equations of turbulent fluctuation. *Quart. Appl. Math.* **3**, 38.

CHOU, P. Y. (1948). On velocity correlations and the equations of turbulent vorticity fluctuation. *Sci. Rep. Nat. Tsing Hua Univ.* **5**, 1.

CORRSIN, S. (1947). Extended applications of the hot-wire anemometer. *Rev. Sci. Instrum.* **18**, 469.

CORRSIN, S. (1949). An experimental verification of local isotropy. *J. Aero. Sci.* **16**, 757.

Corrsin, S. (1951*a*). The decay of isotropic temperature fluctuations in isotropic turbulence. *J. Aero. Sci.* **18**, 417.

Corrsin, S. (1951*b*). On the spectrum of isotropic temperature fluctuations in an isotropic turbulence. *J. Appl. Phys.* **22**, 469.

Corrsin, S. and Uberoi, M. S. (1950). Spectra and diffusion in a round turbulent jet. *Tech. Note, Nat. Adv. Comm. Aero., Wash.*, no. 2124.

Cumming, B. L. (1946). A review of turbulence theories. *Rep. Aust. Coun. Aero.* ACA-27.

Darrieus, G. (1938). Contribution à l'analyse de la turbulence en tourbillons cellulaires. *Proc. 5th Int. Congr. Appl. Mech.* p. 422.

Djang, G.-F. (1948). A kinetic theory of turbulence. *Chin. J. Phys.* **7**, 176.

Dryden, H. L. (1937). The theory of isotropic turbulence. *J. Aero. Sci.* **4**, 273.

Dryden, H. L. (1938). Turbulence investigations at the National Bureau of Standards. *Proc. 5th Int. Congr. Appl. Mech.* p. 366.

Dryden, H. L. (1939). Turbulence and diffusion. *Industr. Engng Chem.* **31**, 416.

Dryden, H. L. (1941). Isotropic turbulence in theory and experiment. *T. von Kármán Anniv. Vol.*, p. 85.

Dryden, H. L. (1943). A review of the statistical theory of turbulence. *Quart. Appl. Math.* **1**, 7.

Dryden, H. L. and Schubauer, G. B. (1947). The use of damping screens for the reduction of wind tunnel turbulence. *J. Aero. Sci.* **14**, 221.

Dryden, H. L., Schubauer, G. B., Mock, W. C. and Skramstad, H. K. (1937). Measurements of intensity and scale of wind tunnel turbulence and their relation to the critical Reynolds number of spheres. *Tech. Rep., Nat. Adv. Comm. Aero., Wash.*, no. 581.

Ellison, T. H. (1951). The propagation of sound waves through a medium with very small random variations in refractive index. *J. Atmo. Terrestr. Phys.* **2**, 14.

Emmons, H. W. (1947). The numerical solution of the turbulence problem. *Proc. 1st Symp. Appl. Math.* Amer. Math. Soc., p. 67.

Favre, A. (1948). Mesures statistiques de la corrélation dans le temps. *Proc. 7th Int. Congr. Appl. Mech.* **2**, 44.

Ferrari, C. (1938). On the theory of turbulence. *Atti Accad. Sci. Torino*, **73**, 373.

Ferrari, C. (1942). The spectrum of turbulence and the statistical theory of isotropic turbulence. *Aerotecnica*, **22**.

Frenkiel, F. N. (1946). Études statistiques de la turbulence. *C.R. Acad. Sci., Paris*, **222**, 367, 473.

Frenkiel, F. N. (1948*a*). On the kinematics of turbulence. *J. Aero. Sci.* **15**, 57.

Frenkiel, F. N. (1948*b*). The decay of isotropic turbulence. *J. Appl. Mech.* **15**, 311.

Frenkiel, F. N. (1948*c*). Comparison between theoretical and experimental results on the decay of turbulence. *Proc. 7th Int. Congr. Appl. Mech.* **2**, 112.

Frenkiel, F. N. (1949). On turbulent diffusion. *Rep. U.S. Naval Ord. Lab.* no. 1136, p. 67.

GOLDSTEIN, S. (1951). On the law of decay of homogeneous isotropic turbulence and the theories of the equilibrium and similarity spectra. *Proc. Camb. Phil. Soc.* **47**, 554.

HALL, A. A. (1938). Measurements of the intensity and scale of turbulence. *Rep. Memor., Aero. Res. Comm., Lond.*, no. 1842.

HEISENBERG, W. (1948*a*). Zur statistischen Theorie der Turbulenz. *Z. Phys.* **124**, 628.

HEISENBERG, W. (1948*b*). On the theory of statistical and isotropic turbulence. *Proc. Roy. Soc.* A, **195**, 402.

HOPF, E. (1948). A mathematical example displaying features of turbulence. *Commun. Pure Appl. Math.* **1**, 303.

INOUE, E. (1951). On the turbulent diffusion in the atmosphere. II. *Geophys. Notes, Tokyo Univ.* **3**, no. 33.

INOUE, E. (1950*b*). On the temperature fluctuations in a heated turbulent fluid. *Geophys. Notes, Tokyo Univ.* **3**, no. 34.

INOUE, E. (1950*c*). On the turbulent diffusion in the atmosphere. I. *J. Met. Soc. Japan*, **28**, 441.

INOUE, E. (1951*a*). On the turbulent diffusion in the atmosphere. II. *J. Met. Soc. Japan*, **29**, 32.

KAMPÉ DE FÉRIET, J. (1938). Some recent researches on turbulence. *Proc. 5th Int. Congr. Appl. Mech.* p. 352.

KAMPÉ DE FÉRIET, J. (1939). Les fonctions aléatoires stationnaires et la théorie statistique de la turbulence homogène. *Ann. Soc. Sci. Brux.* **59**, 145.

KAMPÉ DE FÉRIET, J. (1940). The spectrum of turbulence. *J. Aero. Sci.* **7**, 518.

KAMPÉ DE FÉRIET, J. (1948). Le tenseur spectral de la turbulence homogène non isotrope dans un fluide incompressible. *Proc. 7th Int. Congr. Appl. Mech.* introd. vol.

KAMPÉ DE FÉRIET, J. (1949). Spectral tensor of homogeneous turbulence. *Rep. U.S. Naval Ord. Lab.* no. 1136, p. 1.

VON KÁRMÁN, T. (1937*a*). On the statistical theory of turbulence. *Proc. Nat. Acad. Sci., Wash.*, **23**, 98.

VON KÁRMÁN, T. (1937*b*). The fundamentals of the statistical theory of turbulence. *J. Aero. Sci.* **4**, 131.

VON KÁRMÁN, T. (1938). Some remarks on the statistical theory of turbulence. *Proc. 5th Int. Congr. Appl. Mech.* p. 347.

VON KÁRMÁN, T. (1948*a*). Sur la théorie statistique de la turbulence. *C.R. Acad. Sci., Paris*, **226**, 2108.

VON KÁRMÁN, T. (1948*b*). Progress in the statistical theory of turbulence. *Proc. Nat. Acad. Sci., Wash.*, **34**, 530.

VON KÁRMÁN, T. and HOWARTH, L. (1938). On the statistical theory of isotropic turbulence. *Proc. Roy. Soc.* A, **164**, 192.

VON KÁRMÁN, T. and LIN, C. C. (1949). On the concept of similarity in the theory of isotropic turbulence. *Rev. Mod. Phys.* **21**, 516.

VON KÁRMÁN, T. and LIN, C. C. (1951). On the concept of similarity in the theory of isotropic turbulence. *Adv. Appl. Mech.* **2**, 1.

KELLER, L. and FRIEDMANN, A. (1924). Differentialgleichungen für die turbulente Bewegung einer inkompressiblen Flüssigkeit. *Proc. 1st Int. Congr. Appl. Mech.* p. 395.

KOLMOGOROFF, A. N. (1941*a*). The local structure of turbulence in incompressible viscous fluid for very large Reynolds numbers. *C.R. Acad. Sci. U.R.S.S.* **30**, 301.

KOLMOGOROFF, A. N. (1941*b*). On degeneration of isotropic turbulence in an incompressible viscous liquid. *C.R. Acad. Sci. U.R.S.S.* **31**, 538.

KOLMOGOROFF, A. N. (1941*c*). Dissipation of energy in locally isotropic turbulence. *C.R. Acad. Sci. U.R.S.S.* **32**, 16.

KOLMOGOROFF, A. N. (1949). On the disintegration of drops in turbulent flow. *Doklady Akad. Nauk. S.S.S.R.* **66**, 825.

KOVASZNAY, L. S. G. (1948). Spectrum of locally isotropic turbulence. *J. Aero. Soc.* **15**, 745.

KOVASZNAY, L. S. G., UBEROI, M. S. and CORRSIN, S. (1949). The transformation between one- and three-dimensional power spectra for an isotropic scalar fluctuation field. *Phys. Rev.* **76**, 1263.

LAUFER, J. (1950). Some recent measurements in a two-dimensional turbulent channel. *J. Aero. Sci.* **17**, 277.

LEE, T. D. (1950). Note on the coefficient of eddy viscosity in isotropic turbulence. *Phys. Rev.* **77**, 842.

LIEPMANN, H. W. (1949). Application of a theorem on the zeros of stochastic functions to turbulence measurements. *Helv. Phys. Acta*, **22**, 119.

LIEPMANN, H. W., LAUFER, J. and LIEPMANN, K. (1951). On the spectrum of isotropic turbulence. *Tech. Note, Nat. Adv. Comm. Aero., Wash.*, no. 2473.

LIMBER, D. N. (1951). Numerical results for pressure-velocity correlations in homogeneous isotropic turbulence. *Proc. Nat. Acad. Sci., Wash.*, **37**, 230.

LIN, C. C. (1943). On the motion of a pendulum in a turbulent fluid. *Quart. Appl. Math.* **1**, 43.

LIN, C. C. (1947). Remarks on the spectrum of turbulence. *Proc. 1st Symp. in Appl. Math.*, Amer. Math. Soc., p. 81.

LIN, C. C. (1948*a*). On the law of decay and the spectrum of isotropic turbulence. *Proc. 7th Int. Congr. Appl. Mech.* **2**, 127.

LIN, C. C. (1948*b*). Note on the law of decay of isotropic turbulence. *Proc. Nat. Acad. Sci., Wash.*, **34**, 540.

LIN, C. C. (1950). On Taylor's hypothesis in wind tunnel turbulence. *Memor. U.S. Naval Ord. Lab.* no. 10775.

LOITSIANSKY, L. G. (1939). Some basic laws of isotropic turbulent flow. *Rep. Cent. Aero Hydrodyn. Inst.* (*Moscow*), no. 440. (Translated as *Tech. Memor., Nat. Adv. Comm. Aero., Wash.*, no. 1079.)

MACPHAIL, D. C. (1940). An experimental verification of the isotropy of turbulence produced by a grid. *J. Aero. Sci.* **8**, 73.

MILLIONSHTCHIKOV, M. (1939*a*). Decay of homogeneous isotropic turbulence in a viscous incompressible fluid. *C.R. Acad. Sci. U.R.S.S.* **22**, 231.

MILLIONSHTCHIKOV, M. (1939*b*). Decay of turbulence in a wind tunnel. *C.R. Acad. Sci. U.R.S.S.* **22**, 235.

MILLIONSHTCHIKOV, M. (1941*a*). On the theory of homogeneous isotropic turbulence. *C.R. Acad. Sci. U.R.S.S.* **32**, 615.

MILLIONSHTCHIKOV, M. (1941*b*). On the role of third moments in isotropic turbulence. *C.R. Acad. Sci. U.R.S.S.* **32**, 619.

MOYAL, J. E. (1952). The spectra of turbulence in a compressible fluid; eddy turbulence and random noise. *Proc. Camb. Phil. Soc.* **48**, 329.

OBUKHOFF, A. M. (1941). On the distribution of energy in the spectrum of turbulent flow. *C.R. Acad. Sci. U.R.S.S.* **32**, 19, and *Izv. Akad. Nauk. S.S.S.R.*, Ser. Geogr. i. Geofiz., **5**, 453 (translation issued by Min. of Supply, England, as P21109T).

OBUKHOFF, A. M. (1949*a*). Pressure pulsations in a turbulent flow. *Doklady Akad. Nauk. S.S.S.R.* **66**, 17 (translation issued by Min. of Supply, England, as P21452T).

OBUKHOFF, A. M. (1949*b*). Structure of the temperature field in turbulent flow. *Izv. Akad. Nauk. S.S.S.R.*, Ser. Geogr. i Geofiz., **13**, 58.

OBUKHOFF, A. M. (1949*c*). The local structure of atmospheric turbulence. *Doklady Akad. Nauk. S.S.S.R.* **67**, 643.

OBUKHOFF, A. M. and YAGLOM, A. M. (1951). The microstructure of turbulent flow. *Prikl. Mat. Mekh.* **15**, 3.

ONSAGER, L. (1945). The distribution of energy in turbulence (abstract only). *Phys. Rev.* **68**, 286.

ONSAGER, L. (1949). Statistical hydrodynamics. *Nuovo Cim.*, Supplement, **6**, no. 2, p. 279.

PEKERIS, C. L. (1940). On the statistical theory of turbulence. *J. Aero. Sci.* **8**, 476.

PRANDTL, L. (1938). Beitrag zum Turbulenz Symposium. *Proc. 5th Int. Congr. Appl. Mech.* p. 340.

PROUDMAN, I. (1951). A comparison of Heisenberg's spectrum of turbulence with experiment. *Proc. Camb. Phil. Soc.* **47**, 158.

REISSNER, E. (1938). Note on the statistical theory of turbulence. *Proc. 5th Int. Congr. Appl. Mech.* p. 359.

RICHARDSON, L. F. (1926). Atmospheric diffusion shown on a distance-neighbour graph. *Proc. Roy. Soc.* A, **110**, 709.

ROBERTSON, H. P. (1940). The invariant theory of isotropic turbulence. *Proc. Camb. Phil. Soc.* **36**, 209.

ROTTA, J. (1949). Neue Rechnungen zur statistischen isotropen Turbulenz. *Z. angew. Math. Mech.* **29**, 12.

ROTTA, J. (1950). Das Spektrum isotroper Turbulenz im statistischen Gleichgewicht. *Ingen.-Arch.* **18**, 60.

SATO, H. (1951). Decay of spectral components in isotropic turbulence. *J. Appl. Phys.* **22**, 525.

SCHUBAUER, G. B. (1935). A turbulence indicator utilizing the diffusion of heat. *Tech. Rep., Nat. Adv. Comm. Aero., Wash.*, no. 524.

SCHUBAUER, G. B., SPANGENBERG, W. G. and KLEBANOFF, P. S. (1950). Aerodynamic characteristics of damping screens. *Tech. Note, Nat. Adv. Comm. Aero., Wash.*, no. 2001.

SEDOV, L. I. (1944). Decay of isotropic turbulent motions of an incompressible fluid. *C.R. Acad. Sci. U.R.S.S.* **42**, 116.

SEN, N. R. (1951). On Heisenberg's spectrum of turbulence. *Bull. Calcutta Math. Soc.* **43**, 1.

SIMMONS, L. F. G. and SALTER, C. (1934). Experimental investigation and analysis of the velocity variations in turbulent flow. *Proc. Roy. Soc.* A, **145**, 212.

SIMMONS, L. F. G. and SALTER, C. (1938). An experimental determination of the spectrum of turbulence. *Proc. Roy. Soc.* A, **165**, 73.

STEWART, R. W. (1951). Triple velocity correlations in isotropic turbulence. *Proc. Camb. Phil. Soc.* **47**, 146.

STEWART, R. W. and TOWNSEND, A. A. (1951). Similarity and self-preservation in isotropic turbulence. *Philos. Trans.* A, **243**, 359.

SYNGE, J. L. and LIN, C. C. (1943). On a statistical model of isotropic turbulence. *Trans. Roy. Soc. Can.* section 3, **37**, 45.

TAYLOR, G. I. (1921). Diffusion by continuous movements. *Proc. Lond. Math. Soc.* **20**, 196.

TAYLOR, G. I. (1935*a*). Statistical theory of turbulence. Parts 1–4. *Proc. Roy. Soc.* A, **151**, 421.

TAYLOR, G. I. (1935*b*). Turbulence in a contracting stream. *Z. angew. Math. Mech.* **15**, 91.

TAYLOR, G. I. (1936). Statistical theory of turbulence. Part 5. *Proc. Roy. Soc.* A, **156**, 307.

TAYLOR, G. I. (1937). The statistical theory of isotropic turbulence. *J. Aero. Sci.* **4**, 311.

TAYLOR, G. I. (1938*a*). Production and dissipation of vorticity in a turbulent fluid. *Proc. Roy. Soc.* A, **164**, 15.

TAYLOR, G. I. (1938*b*). The spectrum of turbulence. *Proc. Roy. Soc.* A, **164**, 476.

TAYLOR, G. I. (1938*c*). Some recent developments in the study of turbulence. *Proc. 5th Int. Congr. Appl. Mech.* p. 294.

TAYLOR, G. I. and BATCHELOR, G. K. (1949). The effect of wire gauze on small disturbances in a uniform stream. *Quart. J. Mech. Appl. Math.* **2**, 1.

TAYLOR, G. I. and GREEN, A. E. (1937). Mechanism of the production of small eddies from large ones. *Proc. Roy. Soc.* A, **158**, 499.

TOLLMIEN, W. and SCHAFER, M. (1941). On the theory of wind tunnel turbulence. *Z. angew. Math. Mech.* **21**, 1.

TOWNSEND, A. A. (1947). The measurement of double and triple correlation derivatives in isotropic turbulence. *Proc. Camb. Phil. Soc.* **43**, 560.

TOWNSEND, A. A. (1948*a*). Experimental evidence for the theory of local isotropy. *Proc. Camb. Phil. Soc.* **44**, 560.

TOWNSEND, A. A. (1948*b*). Local isotropy in the turbulent wake of a cylinder. *Aust. J. Sci. Res.* **1**, 161.

TOWNSEND, A. A. (1951*a*). The passage of turbulence through wire gauzes. *Quart. J. Mech. Appl. Math.* **4**, 308.

TOWNSEND, A. A. (1951*b*). On the fine-scale structure of turbulence. *Proc. Roy. Soc.* A, **208**, 534.

TOWNSEND, A. A. (1951*c*). The diffusion of heat spots in isotropic turbulence. *Proc. Roy. Soc.* A, **209**, 418.

VON WEIZSÄCKER, C. F. (1948). Das Spektrum der Turbulenz bei grossen Reynolds'schen Zahlen. *Z. Phys.* **124**, 614.

WIEGHARDT, K. (1941). Comprehensive survey of research on the statistical theory of turbulence. *Luftfahrtforsch.* **18**, 1.

WIENER, N. (1938). The use of statistical theory in the study of turbulence. *Proc. 5th Int. Congr. Appl. Mech.* p. 356.

YAGLOM, A. M. (1948). Homogeneous and isotropic turbulence in a viscous compressible fluid. *Izv. Akad. Nauk. S.S.S.R.*, Ser. Geogr. i. Geofiz., **12**, 501.

YAGLOM, A. M. (1949*a*). On the acceleration field in a turbulent flow *Doklady Akad. Nauk. S.S.S.R.* **67**, 795.

YAGLOM, A. M. (1949*b*). On the local structure of the temperature field in a turbulent fluid. *Doklady Akad. Nauk. S.S.S.R.* **69**, 743.

(*Note added in proof*) The following papers have been published since the Bibliography was prepared:

BURGERS, J. M. (1951). Sur un modèle simplifié de la turbulence. *Publ. Sci. Tech. Ministère l'Air*, no. 251.

CHANDRASEKHAR, S. (1951*c*). The fluctuations in density in isotropic turbulence. *Proc. Roy. Soc.* A, **210**, 18.

COBURN, N. (1950). A method for constructing correlation tensors in homogeneous turbulence. *Proc. 1st Midwest Conf. Fluid Dynamics, Ann Arbor*, p. 129.

HOPF, E. (1952). Statistical hydromechanics and functional calculus. *J. Rat. Mech. Anal.* **1**, 87.

INOUE, E. (1951*b*). Some remarks on the dynamical and thermal structure of a heated fluid. *J. Phys. Soc. Japan*, **6**, 392.

INOUE, E. (1951*c*). The application of the turbulence theory to the large-scale atmospheric phenomena. *Geophys. Mag., Tokyo*, **23**, 1.

KAMPÉ DE FÉRIET, J. and BETCHOV, R. (1951). Theoretical and experimental averages of turbulent functions. *Proc. Acad. Sci. Amst.* **54**, 389.

MATTIOLI, E. (1952). La teoria statistica della turbolenza. *Aerotecnica*, **32**, 3.

PROUDMAN, I. (1952). The generation of noise by isotropic turbulence. *Proc. Roy. Soc.* A, **214**, 119.

UBEROI, M. S. and CORRSIN, S. (1952). Diffusion of heat from a line source in isotropic turbulence. *Tech. Note, Nat. Adv. Comm. Aero., Wash.*, no. 2710.

(*Note added in* 1956.) A recent paper which modifies some of the developments described in §§ 3.1, 5.3 and 5.4 is:

BATCHELOR, G. K. and PROUDMAN, I. (1956). The large-scale structure of homogeneous turbulence. *Philos. Trans.* A, **248**, 369.

INDEX